STEAM ALONG THE BOUNDARY

Canadian Pacific, Great Northern and the Great Boundary Copper Boom

Robert D. Turner and J. S. David Wilkie

sononis
PRESS
WINLAW, BRITISH COLUMBIA

Cover Painting...
A long Canadian Pacific ore train eases over the truss bridges and trestles that spanned the North Fork of the Kettle River at the power dam near Grand Forks' Granby Smelter. Canadian Pacific began serving the smelter with ore from the Phoenix mines by mid-1900. The rather frail-looking wooden truss spans were replaced by a steel bridge on masonry piers and abutments in 1914. In February 1905, the Great Northern completed its own branch line to the Granby Smelter including another crossing of the river just upstream from the CPR bridge.
–PAINTING BY MAX JACQUIARD © 2006.

Dedication

In Memory of Dave Wilkie

A fine photographer, distinguished railway historian, and generous friend.

In 1996 Dave won the Canadian Railroad Historical Association's Lifetime Achievement Award; it was well deserved.

Photos, facing page...
The copper smelter at Greenwood was one of the major industrial operations along the boundary and processed ore from the Mother Lode Mine. A Canadian Pacific Shay locomotive is arriving at the smelter from the mine in this photo from about 1901. A larger brick smokestack that still stands replaced the one in the photo and the smelter was expanded considerably. In the lower photo, CPR Nos. 5125 and 3647 on Train No. 12, the *Kettle Valley Express*, eastbound near Nelson in 1952.
–BC ARCHIVES, 9994, ABOVE, AND W. GIBSON KENNEDY, GORDON KENNEDY COLLECTION, WITH THANKS TO LANCE CAMP, BELOW.

Library and Archives Canada Cataloguing in Publication

Turner, Robert D., 1947-
Steam along the boundary : Canadian Pacific, Great Northern and the great boundary copper boom / Robert D. Turner, J.S. David Wilkie.

Includes bibliographical references and index.
ISBN 978 1-55039-158-9

1. Canadian Pacific Railway Company—History. 2. Great Northern Railway Company (U.S.)—History. 3. Copper mines and mining—British Columbia—History. 4. Copper mines and mining—Washington (State)—History. I. Wilkie, J. S. David II. Title.

HE2701.T87 2007 385.09711'62 C2007-901218-3

Sono Nis Press most gratefully acknowledges support for our publishing program provided by the Government of Canada through the Book Publishing Industry Development Program (BPIDP) and the Canada Council for the Arts, and by the Province of British Columbia through the British Columbia Arts Council and the Book Publishing Tax Credit, Ministry of Provincial Revenue.

Copy edited by Dawn Loewen
Designed by Robert D. Turner

Published by
SONO NIS PRESS
Box 160
Winlaw, BC V0G 2J0
1-800-370-5228

books@sononis.com
www.sononis.com

Printed and bound in Canada
by Friesens Printing

The Canada Council for the Arts | Le Conseil des Arts du Canada

Contents

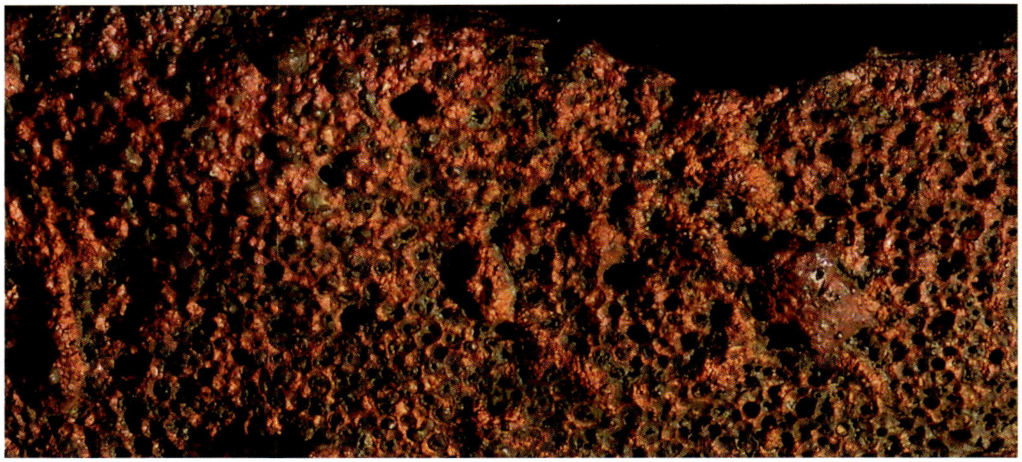

A Wealth of Copper...

Phoenix meant copper. And with it smelters, railways and cities. The copper-silver-gold ore, like this sample above from the Old Ironsides Mine, yielded the Boundary's blister copper. Railways took the blister to refineries to yield the pure copper used for thousands of products. But by the 1920s it was over and Phoenix arose from its ashes only long enough, decades later to be mined out and to disappear forever.

–ORE SAMPLE AND BLISTER COPPER, ROYAL BC MUSEUM, ROBERT TURNER PHOTOS; OLD IRONSIDES, DAVE WILKIE COLLECTION; SMELTER, MCRAE PHOTO, WILSON COLLECTION, OKANAGAN ARCHIVES TRUST; ENGINE CREW, GREENWOOD MUSEUM; AND PHOENIX PIT, DAVE WILKIE.

Now Only Memories...

In just over 20 years the boom was over, the mines and smelters closed. A century later all that remained were the slag dumps, Greenwood's looking like some strange battle-field strewn with the helmets of giants, a few bridge abutments, and abandoned railway routes such as the CPR's line high above Christina Lake, shown in July 1978.
–Robert D. Turner.

Lorraine and Mike Wilkie on the abandoned VV&E grade west of Myncaster, B.C., at mile 83.6 to be precise, on October 9, 1966. The Wilkie family explored many miles of long-abandoned railways.

–Dave Wilkie.

Sarah Turner watches Canadian Pacific GP9 diesels with only a caboose in tow, rolling through Cascade, B.C., on a very hot and beautiful July day in 1978. Ice cream awaited in Grand Forks for all of us.

–Robert D. Turner.

Columbia Legislative Library, Victoria; California State Railroad Museum, Sacramento; Canadian Museum of Rail Travel, Cranbrook; Canadian Pacific Archives, Montreal; Canadian Pacific Railway's engineering and public relations offices in Vancouver and Calgary; Ferry County Museum, Republic; Glenbow Museum, Calgary; Great Northern Railway, Public Relations Department, in particular Frank Perrin, St. Paul; Great Northern Railway Historical Society, St. Paul; Greenwood Museum of the Greenwood Heritage Society, Greenwood; James J. Hill Library, St. Paul; Kelowna Museum, Kelowna; Kettle River Museum Society, Midway, B.C.; Kootenay Museum Association, Nelson; Minnesota Historical Society, St. Paul; Museum of History and Industry, Seattle; National Archives of Canada, Ottawa; Northwest Museum of Arts & Culture/Eastern Washington State Historical Society, Spokane; Okanagan Archives Trust Society, Penticton; Old Depot Historical Museum, Oroville; Penticton Museum & Archives, Penticton; Princeton and District Museum and Archives, Princeton; Rossland Museum and Archives, Rossland; Royal British Columbia Museum, Victoria; Spokane Public Library, Spokane; Trail Museum & Archives, Trail; University of British Columbia Library, Special Collections, Vancouver; University of Washington Library, Special Collections, Seattle; City of Vancouver Archives, Vancouver; and the Vancouver Public Library, Vancouver. A special thanks to Barbara McLellan and Peter Westerby, both formerly with the Provincial Archives, for their help in finding and printing photos for us many years ago. We also benefited from the talents of Rob and Sarah Tyler at Island Transparency Service, Bob Mattheson, Jack Fawcett and our friends at Crystal Finish Photo Service, and Chu Yeung and Lee Prouting who made many of Lance Camp's beautiful prints, working patiently with aging negatives.

Over the years our families were often part of our fieldwork during vacations, making the explorations enjoyable, and helping in many other ways. Special thanks go to Dave's late wife, Bea, and to Michael Dunham-Wilkie and Lorraine Wilkie Buchanan. Moreover, Lorraine spent hundreds of hours sorting Dave's papers and making them available to me for this book. Dave wanted to extend his warm appreciation to Anne Curson, his companion and special friend during his illness.

Dave's family, concerned that his wonderful photos and research material continue to be available, graciously donated his collection to the West Coast Railway Association at Squamish, where it will be available for researchers in the future and contribute to many projects, just as Dave himself would have done so generously.

My family contributed to this story in countless ways, spending many holidays in the southern interior, exploring around Greenwood, Princeton, Grand Forks, Midway and other places important to this history—to Nancy, Sarah, Molly and Kate Turner my sincere thanks. Nancy added many insights to the history and ecology of the region as well as being a skilled editor, and a most sympathetic, warmly encouraging and insightful counsel when I needed it most. My mother, Isabella, deserves a special thanks for her help on numerous trips and endless encouragement for my writing projects.

I also want to thank Max Jacquiard, that talented railway artist and good friend, whose beautiful painting of a Canadian Pacific ore train from Phoenix is on the cover of this book. A profile of Max is on the next page.

A manuscript does not become a book without a publisher and I am privileged once again to work with my friends of many years at Sono Nis Press. My sincere thanks are due Jim Brennan for his production assistance, Dawn Loewen for her excellent suggestions and copy editing, and especially to my friend and publisher Diane Morriss for her enthusiasm and continuing support. Rarely does an author have the pleasure of working with three generations, actually four, of the same family with such continuing respect, good humour, trust and friendship. I hope you enjoy *Steam Along the Boundary* and that the "old timers," our respected cultural elders, will feel it has told their stories well and with respect.

Robert D. Turner
Victoria, B.C.

November 7, 2006

Max Jacquiard and the Cover Painting

Phoenix Ore for the Grand Forks Smelter

Max Jacquiard is the talented artist who graciously provided the painting of a Canadian Pacific ore train from the mines at Phoenix, B.C., for the cover of this book. The scene shows the train easing over the North Fork of the Kettle River as it nears the smelter at Grand Forks.

Max is well known for his artistic and accurate renderings of railways in Canada and in particular for breathtaking scenes in British Columbia and the Rockies. Born in Flin Flon, Manitoba, in 1934, Max moved to the coast ten years later. "I liked painting scenes and portraits," he recalls, "then it was like a light coming on. I loved trains when I was a kid and I wanted to paint them from the days I remembered as a boy." He works mostly with acrylics on canvas, now doing eight to ten large paintings a year. Canadian Pacific scenes predominate, because, he concludes, "it has so many interesting places. I've done lots at Banff and in the Rockies. CN ran along the valleys and didn't have the pusher districts in the mountains, but it also had a lot of beautiful sections and ran through Jasper and the Thompson Canyon." Field, on the western side of the Rockies, is a favourite area as are the lower stretches of the Thompson River east of Lytton, where Canadian Pacific and Canadian National, on opposite sides of the river, run through deep and spectacular canyons in a semi-arid landscape. Light and shadow play an important role in Max's work and he paints scenes he knows first-hand from extensive travels along the routes through the mountains and canyons. Many of his paintings are in private collections, and limited edition prints have been published since 1993 by Banman Publishing of Coquitlam, B.C. The photo shows Max with his painting of a CPR train passing the old church at Notch Hill, along the busy CPR main line between Sicamous and Salmon Arm.

One hundred years after the train depicted in Max's cover painting rumbled across the smelter bridge, only the bridge piers of a later railway bridge, the overgrown roadbed and some pilings remain at this pretty spot along the river. Now it is hard to imagine, standing above the North Fork of the Kettle, how busy and industrial this peaceful landscape was just a few generations ago.

Tangible reminders endure of the Boundary railways and the Granby Smelter, such as masonry piers and abutments that survive along the North Fork of the Kettle, now known as the Granby River, near Grand Forks, but it takes insight and imagination through an artist's eye to bring scenes to life–like the CPR's crossing of the Granby at this spot–a century later.
–Robert Turner, August 2004.

Max Jacquiard in his Coquitlam, B.C., studio in 2005, painting the scenes of steam railways in the mountains that he loves best.
–Nancy Turner.

C.P.R. 952

Introduction

Breathtaking and Beautiful Railway...

"The trip from Kaslo to Greenwood was one of grandeur unexcelled. From the long sloping hills around Robson to the sharp peaks at Phoenix one would hardly think that the ingenuity of man could accomplish so much in a few months.... From Robson to the Summit is a steady climb of many feet and as you go through a tunnel—of which there are many—and onto a high trestle, it almost takes your breath away. ...At the summit there are nine switchbacks of half a mile each going up, and the same on the other side coming down. It takes two hours to go through all this maneuvering and then you start on the way to the famous Boundary Country."

—*The Kootenaian* (Kaslo), December 14, 1899.

The Premier Copper Producing District in Canada...

"The Boundary District entered the ranks of the copper producing centres of British Columbia in 1900, and ever since has held the leading place in the Province, and for some years has been the premier copper producing district in Canada.

"The mines, up to July 1, 1910, have produced and shipped over 7,000,000 tons of ore."

—O. E. LeRoy, *The Geology and Ore Deposits of Phoenix,* 1912.

The interplay of mines, smelters and railways set the stage in the Boundary District for a 20-year mining boom that shaped the formative years of many cities and communities, some now all but forgotten, along the British Columbia–Washington border. At left, Canadian Pacific's locomotive 952, a 2-8-0, built in 1896 and first used on the Field Hill in the Rockies, paused for this 1903 Notman photo at the Mother Lode Mine near Greenwood. The Boundary District railways received some of the CPR's newest locomotives in the early 1900s because of the demanding nature of the traffic and the steep grades on the branch lines to the mines. This one, later renumbered 3131, ended its days on Canadian Pacific's Esquimalt & Nanaimo Railway on Vancouver Island in the early 1930s.
–Wm. Notman & Son Photo, McCord Museum, v3684.

Copper brought electricity to the industrial and urban world of the late 1800s and early 1900s. Coal fuelled the Industrial Revolution, and iron and steel built it, but copper, because of its many industrial applications and, critically, its ability to conduct electricity over long distances, brought with it riches and a revolution just as important and dramatic.

Usually found with traces of gold and silver, copper ore was one of the keys to a modern and prosperous industrial economy and it was also crucial in armaments for everything from shell casings to the wiring in a Dreadnought battleship. In the tumultuous years of the early 1900s, copper production on a world scale increased dramatically and its mining and smelting became critically important along the southern boundary of British Columbia and northern Washington State. For a few short decades the mines and smelters of the region pulsed with activity and drove a burgeoning economy, but by the early 1920s, they were all but gone. *Steam Along the Boundary* presents the story of these mines and the railways that served the area.

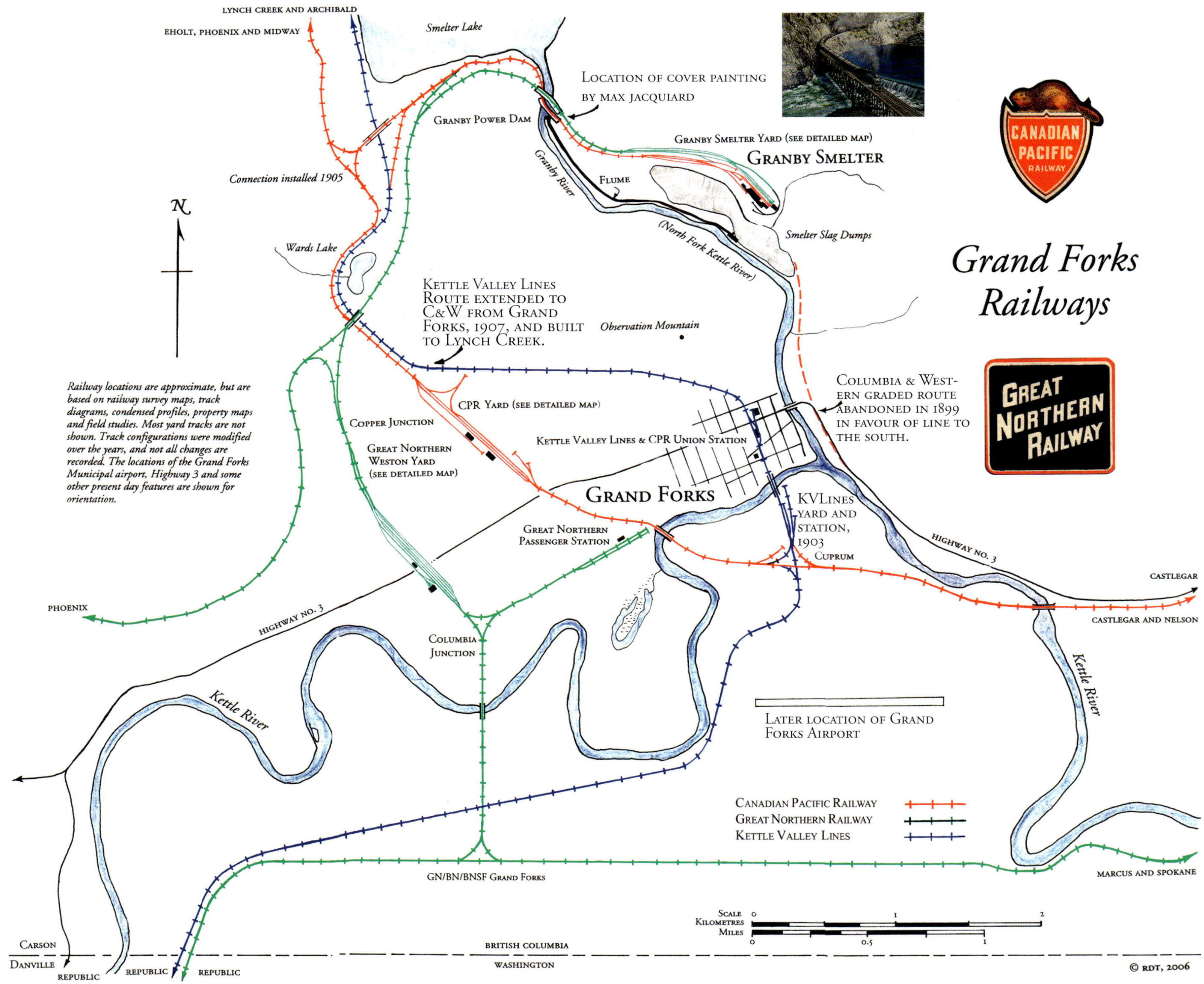
Grand Forks Railways
LYNCH CREEK AND ARCHIBALD
EHOLT, PHOENIX AND MIDWAY
Smelter Lake
Location of cover painting by Max Jacquiard
Granby Power Dam
Granby Smelter Yard (see detailed map)
Granby Smelter
Canadian Pacific Railway
Connection installed 1905
Granby River
Flume
Smelter Slag Dumps
(North Fork Kettle River)
N
Wards Lake
Kettle Valley Lines route extended to C&W from Grand Forks, 1907, and built to Lynch Creek.
Observation Mountain
Railway locations are approximate, but are based on railway survey maps, track diagrams, condensed profiles, property maps and field studies. Most yard tracks are not shown. Track configurations were modified over the years, and not all changes are recorded. The locations of the Grand Forks Municipal airport, Highway 3 and some other present day features are shown for orientation.
Columbia & Western graded route abandoned in 1899 in favour of line to the south.
Great Northern Railway
CPR Yard (see detailed map)
Copper Junction
Kettle Valley Lines & CPR Union Station
Great Northern Weston Yard (see detailed map)
Grand Forks
KVLines yard and station, 1903
Great Northern Passenger Station
Cuprum
Highway No. 3
Castlegar
Castlegar and Nelson
Phoenix
Highway No. 3
Columbia Junction
Kettle River
Kettle River
Later location of Grand Forks Airport
Canadian Pacific Railway
Great Northern Railway
Kettle Valley Lines
GN/BN/BNSF Grand Forks
Marcus and Spokane
Scale
Kilometres
Miles
0
1
2
0
0.5
1
Carson
Danville
Republic
Republic
Republic
British Columbia
Washington
© RDT, 2006

The Boundary District of British Columbia and the adjacent parts of Whatcom, Okanogan and Ferry counties in Washington State are rich in history but to most people travelling through these scenic areas, the nature of what drove the development of towns and cities in the region over a hundred years ago is little known or appreciated. Few people can imagine the boom years of now quiet communities such as Greenwood or Republic. Casual travellers might think that Greenwood, for example, is little changed from those days because of its fine representation of quaint and elegant buildings from the late 19th century, but they are a legacy of an industrial era that painted a very different picture. It was a time of bustling railways, with coal and coke trains from the Crowsnest Pass,[1] crowded branch line passenger trains, ore trains from the Mother Lode copper mines and locomotives endlessly switching the industrial trackage around the mines and smelters. There was a huge copper smelter just across the valley from Greenwood, working 24 hours a day, its acrid fumes often filling the valley, the huge fiery slag piles aglow at night as trains dumped molten waste hour after hour, while throughout the day whistles were blowing at the power plant, and the sounds of commerce and rowdiness echoed along the busy streets, all confined to the narrow valley of Boundary Creek. And there was another copper smelter just a few miles south at Boundary Falls. At the beginning of the 20th century, Greenwood's population of more than 3,000 was five times what it was a century later when just 600–700 people lived in the area, with quite a few of those residents living there in retirement because of comparatively low-cost housing.

This region experienced a dramatic and fast-paced, mining-based, railway-building boom in the late 1800s and early 1900s. Hot on the heels of the mining excitement that brought international attention to Rossland, the Slocan and other districts not far to the east, the Boundary arose as one of the potentially most productive and rich areas yet being developed, and a focus of two highly competitive and expanding North American transportation giants: the Canadian Pacific Railway and the Great Northern Railway. The story that developed around these railways, the mines and smelters that became critical to their survival, the intertwined aspirations of ambitious individuals, and the corporate politics they provoked produced one of the most fascinating chapters in the transportation and industrial history of British Columbia and Washington.

Central to this period is the image of Canadian-born James Jerome Hill, president of the Great Northern Railway, railroad baron, industrial giant, and one of the most imposing and influential figures of his time, sometimes called the "gilded age" in American history. Also a key figure in the years preceding the Canadian Pacific's construction to the Boundary District mines was Hill's one-time colleague and friend, then corporate adversary and business rival, American-born William Cornelius Van Horne, president of the Canadian Pacific Railway. Van Horne, who was knighted in 1894, retired as Canadian Pacific president in 1899 as the CPR was expanding westward to the Boundary District, and became chairman of the Canadian Pacific board, a position he held until 1915. It fell to Van Horne's successor as president, the extremely capable Thomas George Shaughnessy, to oversee the development of the railway through southern British Columbia in the early 1900s as the corporate rivalries continued. The Great Northern story also features Hill's somewhat more reserved yet highly competent elder son, Louis Warren Hill, who became GN president in 1907 when his father became board chairman. Louis Hill, like Thomas Shaughnessy, was far less flamboyant than his predecessor. They both proved to be excellent managers and consolidators of the vast corporate empires they controlled. They were financially astute, complex, self-assured, demanding and ambitious, and their highly successful railway companies in many ways reflected their personalities and talents.[2]

The Great Northern's route across northern Washington, built as the Washington & Great Northern Railway, and through southern B.C., with the expansive title of the Vancouver, Victoria & Eastern Railway & Navigation Company, was a twisting, turning, climb-

William Cornelius Van Horne, later Sir William, was a man of drive, ambition and determination. Like the other railway barons of his era, including James J. Hill, his convictions and ego matched the corporate empire he developed. He made his reputation during the construction of the main line of the Canadian Pacific and went on to be its president. Van Horne, like his highly capable successor, Thomas Shaughnessy, was fiercely determined to expand and protect the CPR and maintain what he saw as Canadian interests. "Wisely or not, the Canadian Pacific Railway was created for the purpose of developing the Great West of Canada and for reaching out for trade beyond in the interest of the country at large...," he wrote the Toronto *Globe's* editor. *"Every line permitted to enter Canadian territory from the south is a weapon in the hands of the competitors of the Canadian Pacific."* [emphasis Van Horne's] –CANADIAN PACIFIC ARCHIVES.

Looking every bit the stern patriarch that he was in this photo from his later years, James Jerome Hill was the driving force behind the Great Northern Railway and a formidable corporate opponent in the struggles between the major railways for influence, traffic and expansion that reached their height in the years leading up to the First World War. Hill, often called the Empire Builder, played out his great game of railway chess across the Northwest states and southwestern Canada. Canadian born, he was a long-term associate and advisor of the CPR, having once been on its board. Writing privately, Van Horne, who played the game equally well, once described Hill as the "bitterest enemy Canadian interests have in the United States, and the most dangerous because the most unscrupulous." Nonetheless, they remained cordial as they jostled for position.
–Boundary Museum, 987.129.004.2.

ing line that crossed the international boundary five times between Kettle Falls, Washington, and Princeton, B.C., and if the two short spurs are counted (the "Chesaw Spur" near Myncaster that did not actually reach Chesaw, and the Sidley spur just west of Molson) the boundary between Midway and Molson are counted, one could say there were seven border crossings. Following an easy grade much of the way, except between Molson and Oroville, the GN line ran north and south as much, it seemed, as it ran east and west, but for all its roundabout qualities it still maintained a less arduous route than the Canadian Pacific's trackage between the Columbia River at Castlegar and its entry into the mining town of Princeton far to the west. One of Hill's fundamental notions about railroad construction and operation was to build the best rail line possible with the gentlest available grades and least curvature. If the railroad could be operated efficiently, he reasoned, it could handle tonnage cheaply and effectively and this would justify the greater initial investments in right-of-way construction. His idea proved sound and it gave his Great Northern a competitive advantage over other railways for the long-term traffic that usually meant profitable operations. Canadian Pacific's route was constrained by the decisions that restricted it from crossing and recrossing the international boundary and not being able to follow the rivers and easier alignments that might have been possible. As a result, it had steeper gradients and more mountain crossings as it snaked its way along the Canadian side of the international boundary.

The Great Northern withdrew from most of the mining districts of British Columbia long before the CPR, but ironically, in the end, it was Great Northern trackage to Grand Forks and Curlew on the once busy route to Republic that endured longest. It continued to serve parts of the region on both sides of the border into the 21st century, many decades after the mining boom ended, and when the bustling days of ore trains and mining camps were all but forgotten. At Grand Forks, the Kettle Falls International Railway, owned by OmniTRAX, a shortline railroad company, operates over the former Burlington Northern Santa Fe route along the boundary that now ends just south of the international boundary, north of Republic, and makes a brief jog into Canada to serve Grand Forks before turning south towards Colville and Spokane. BNSF, the corporate descendent of the Great Northern, sold its trackage west of Kettle Falls to OmniTRAX in 2004. Only a small remnant of CPR trackage remains in Grand Forks, now operated by the Grand Forks Railway Company, providing local industries with switching services to the former Burlington Northern Santa Fe's interchange south of the city. All the way west to Princeton, through Bridesville, Molson, Hedley and other towns along the border, there are many traces of what was once the Great Northern's so-called "Third Main Line" that wound its way from the heart of the Inland Empire at Spokane, through southern British Columbia and northern Washington to Vancouver on the coast of British Columbia.

The CPR, once the dominant force in transportation across this vast expanse of mountains, has withdrawn completely to the east. The once busy Kettle Valley Railway and the Boundary Subdivision from Midway, twisting through the mountains to Grand Forks and over Farron Hill to Castlegar on the Columbia River, are now abandoned, the tracks gone and the old grade silent. Along many hillsides and through the farmlands there is still much evidence of the railways, and large sections of the CPR's route are now part of the Trans Canada Trail system. Today you can travel from Hope, at the head of the Fraser Valley, east to Grand Forks before encountering a railway, except for the fine heritage operation of the Kettle Valley Steam Railway at Summerland.

All books must have boundaries and *Steam Along the Boundary* is no exception. It focuses primarily on the history of these two major railways through the Boundary District and adjacent areas in British Columbia and northern Washington. Secondarily, it explores the development of the mining and smelting industries that provided much of the justification and traffic for these railway lines. The mining history is very complex and could be the subject for a book in itself. This book also brings in the story of the Kettle Valley Railway, which was Canadian Pacific's extension westward from Midway through to Hope. Because the Kettle Valley's account is already well documented, its story is limited to only those aspects of its construction and later history that relate to the Great Northern extensions westward, the competitive position of the two railways, the operations of the CPR's other lines in the southern interior and its overall importance to the communities of the Boundary District. We have also, for the sake of space, not treated the CPR branch lines between Castlegar, Trail and Rossland, the history of the railways to the Rossland–Trail mining region and the giant Cominco Smelter at Trail. That area also deserves a book of its own.

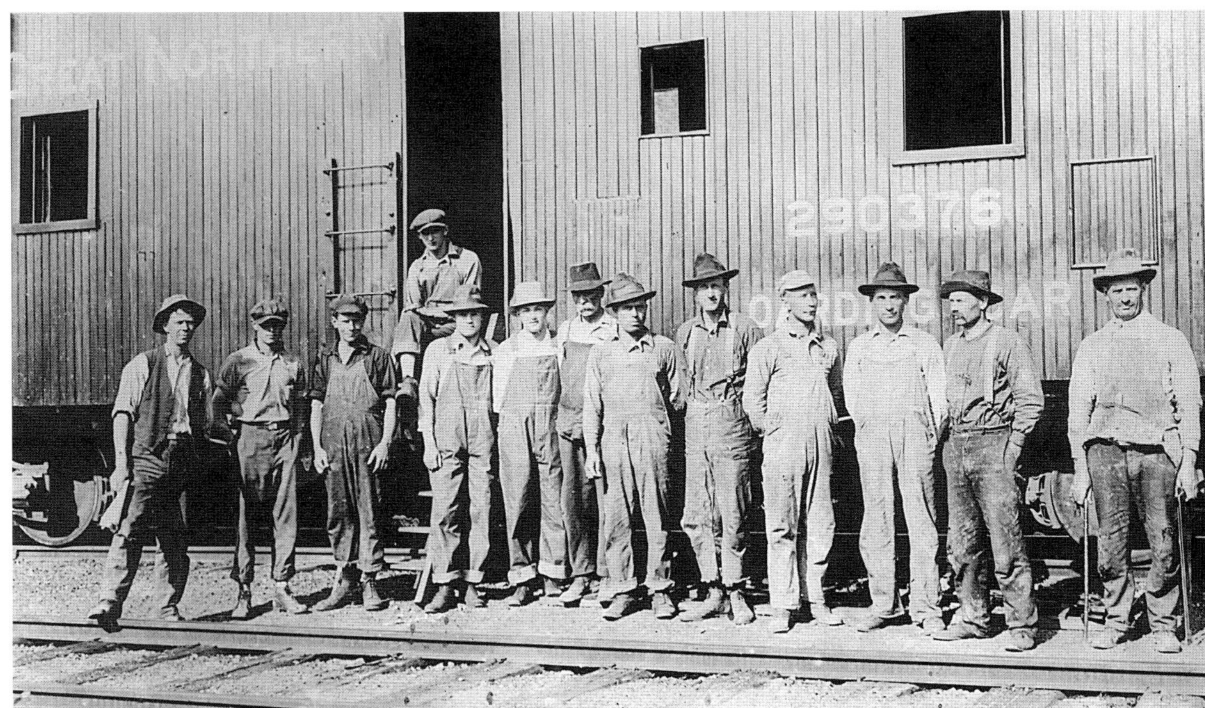

Life in the mining and smelting towns of the industrial frontier was hard and offered few amenities, but the residents soon found opportunities for social events and other activities to make their lives more pleasant and enjoyable. Excursions by train were always popular, as the turnout for this Knights of Pythias special suggests. Such trips to neighbouring communities for special events, sports activities or picnics provided chances for wearing your new best dress or suit, socializing, gossip, meeting other young people in appropriate circumstances and a change of scenery.

For construction workers, like these men at right, outside a Great Northern boarding car, conditions could be rugged. The transient, virtually male-only camps sometimes had cars such as these, but often the men lived in tents with few comforts, frequently in the rain and snow. The hours were long, the food questionable, and the company tired, grumpy, needing a bath and clean clothes, and frequently speaking another language. Some labourers came from farms in the fallow season, but most were recent immigrants from Europe. Unlike in the 1880s when the CPR was built through the Fraser Canyon, Asians were not normally employed by the contractors; it was a time of resentment and hostility towards immigrants from China, Japan or India, who were seen as taking away jobs and were unwelcome in railway construction and most other industrial jobs except in some parts of the fishing and canning industries, and to an extent in sawmilling.

–BOUNDARY MUSEUM, 991.055,005.31 AND 002.487.001.

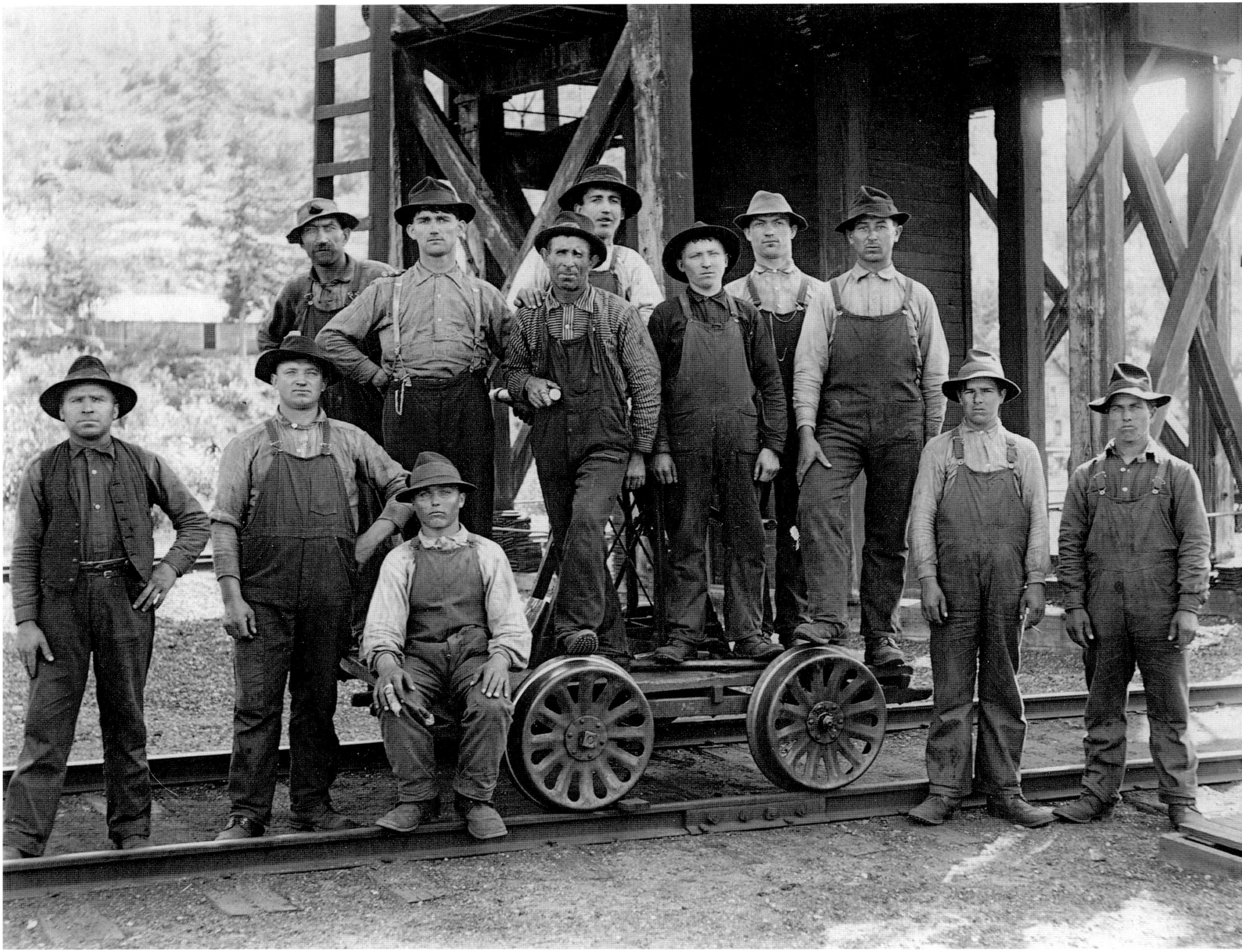

Railway Names, Abbreviations and Charters

Railway companies normally required the granting of a legal charter, or its equivalent, from the government before they could raise funds for construction, acquire property, build their trackage and eventually open for traffic. These charters were often a focus of speculation, and many changed hands or were "fronted" by people on behalf of the larger corporations that might have intimidated or alienated local or regional interests. Moreover, just the idea that a major company might be involved in a charter could send the price of private lands skyrocketing. The large companies welcomed the inflation of property values, but only after they had secured the lands they needed and were ready to sell other properties. A multitude of different charters was proposed, promoted or granted to permit rail line construction in British Columbia and Washington during the late 1800s and early 1900s. The various corporate names and their usual abbreviations can become confusing even to the most dedicated railway historian trying to unravel their history. I have used the proper names to introduce these schemes and companies and where possible simplified their use by referring to the major system that owned or controlled them. For example, in most cases, the Washington & Great Northern, the GN's subsidiary in northern Washington, is called simply the Great Northern. Similarly, the Columbia & Western, which was the CPR's line from West Robson on the Columbia River, just north of Castlegar, to Midway is usually called simply the CPR. The maps in this book provide further details of these components of the major systems.

Of the two major railways that dominate this story, only the Canadian Pacific survives as an operating company. James J. Hill's Great Northern Railway itself is but a memory; the railway was included in the merger of May 2, 1970, that formed Burlington Northern Incorporated, bringing together the Great Northern Railway, the Northern Pacific Railway and the Chicago, Burlington & Quincy, and later the Spokane, Portland & Seattle Railway, Frisco and other related companies. In 1995, BN became part of the even bigger Burlington Northern Santa Fe Corporation that through its subsidiary, the Burlington Northern Santa Fe Railway, operates the former BN trackage through the Northwest and the famous Santa Fe system from Chicago across the southwest states to California. The railway itself is now usually called simply the BNSF Railway. BNSF, like most other major rail companies, has divested itself of branch lines through abandonment, lease or sale, and the trackage from Kettle Falls to the Boundary District was sold in 2004 to the Kettle Falls International Railway owned by OmniTRAX.

For safety reasons, railways in Canada were not required to change their traditional designation of locations along their lines from miles to kilometres. Although distances are converted for the convenience of readers, specific milepost references have been left in original form. The following list provides the names (with abbreviations where used in the text or on maps) of the principal railways involved in the story of the Boundary region and adjacent areas.[1] Some railways that played only a minor role or were Great Northern or Canadian Pacific properties elsewhere are not included.

CANADIAN PACIFIC RAILWAY COMPANY
PACIFIC DIVISION.

The Kettle Valley Railway Company.

Facing page...
History is so often defined and described by great events and the leaders of different times. While leadership is critical to nearly all endeavours, it is people doing their jobs on a day-to-day basis, and supporting their families, that are the real backbone of any enterprise. This could never be truer than when considering the development and operation of the railways, mines and smelters in the frontier settlements of British Columbia and Washington State. Fortunately, the names of most of these sectionmen from Phoenix on the Great Northern in 1901 are known: Foreman Gabriel Palek, Joe Palek, Chas. Palek, Chas. Yetko, Steve Yankola, Andy Yankola, Joe Kravsky and Paul Kravsky. It was quite common for several members of the same family to work together as sectionmen on the railways.

Men who worked on the railways, like these Great Northern employees, were responsible for track maintenance, repairs and re-railing wrecks. It was a hard and uncompromising life; consider that two of these men have bandaged hands reflecting, most probably, the hard routine duties of their jobs–heavy steel, freezing weather, rocks and gravel, timbers and rubble.
–Boundary Museum, 987.034.

Part 1: Rails to the Copper Camps

Canadian Pacific and Great Northern Build to the Boundary

The mining and smelting boom towns that appeared almost overnight, it seemed, in the Boundary District of southern British Columbia and the adjacent parts of northern Washington could not have grown or survived without the extension of the railways to service the mining industry. The railways provided access to refineries and markets, as well as supplies of coal and coke without which the mining industry could not have developed on a large scale. The railways also made it possible for the growing towns to import food and all the necessities and luxuries of life to the mining frontier. Greenwood, shown at left in the early 1900s, was one such city and for 20 years it owed most of its livelihood to the copper mines in the nearby mountains and the huge smelter that was built just south of town at a site called Anaconda.

The mining boom required new equipment to handle the heavy ore tonnages coming from the mines. Canadian Pacific brought three Shay locomotives, for a time their most powerful engines, to Phoenix and the Mother Lode.

–Cominco Collection, left; Elwood White Collection, above.

Railway Beginnings…

"That the interests of your Company may be protected in Southern British Columbia, it is necessary to move on westward from the Columbia River at Robson, the western end of your line, so as to reach the Boundary Creek District—about 100 miles—during the present year…"

—Canadian Pacific Railway, Annual Report, 1897.

"A long looked for arrival of a steam locomotive happened last Wednesday. A big engine decorated with Union Jacks and Old Glories that looked as if they had campaigned in Cuba pushing a tracklayer ahead of it. The rails were laid with remarkable rapidity and by 4 o'clock in the afternoon the engine had passed the C.P.R. hospital. Shortly after 4 p.m. another engine came in with a passenger coach. The occupants included Collingwood Schreiber, Deputy Minister of Railways, W.F. Tye…"

—*The Boundary Creek Times*, October 21, 1899.

The first railroad spanning the vast interior of North America to reach the Pacific was the combined route of the Union Pacific and Central Pacific railways completed at Promontory, Utah, in 1869. Railways were prominent in almost any discussion of the West, whether in Canada or the United States, but there were many delays before construction proceeded at a concerted pace on railroads through the northern states to Puget Sound and across the Canadian Prairies to the Pacific coast of British Columbia. The rate of expansion was a slow and frustrating one for the promoters, miners and settlers who filtered in to this vast and beautiful area. Eventually, the enormous and challenging task of surveys, the securing of financing, and all of the political questions that needed resolution were addressed and construction could begin, gathering momentum as the rails were pushed through the mountains. Once work began at last, over a decade of construction would see three major transcontinental lines completed across the Rockies to the Pacific. "Not until a frenzy of track laying gripped the region between 1880 and 1893 did railroads write clearly legible signatures across the Pacific Northwest," notes Carlos Schwantes in his excellent history of railroads in the Northwest states.[1] The building boom stagnated with the economic depression of 1893, but resumed by 1897 as the economy recovered. The Northern Pacific Railway (NP) was the first of the northern transcontinental railways to reach the Pacific when it was completed through

Building a railway was one thing, but building the traffic to cover its costs was an ongoing worry...

CONFIDENTIAL

11th. August, 1899

William Whyte, Esq.,
Manager of Lines West,
Winnipeg, Man.

Dear Sir,

I am very anxious about the development of traffic on the Crow's Nest Line. We have a large amount of money invested there now, and up to the present time the net earnings of the line have been small, as compared with the interest on the capital. No doubt too a considerable portion of the earnings given to the line come out of the revenue of the route via Revelstoke. Are you about to suggest anything that we could do to encourage the establishment of industries or some other way to put the line on a more profitable basis?

Yours truly,

T. G. Shaughnessy
President

–SHAUGHNESSY PAPERS, CANADIAN PACIFIC ARCHIVES, FILE 518.

to Tacoma in 1883, two years ahead of the Canadian Pacific. The NP's route was at first rather roundabout and followed the Columbia River to Kalama before swinging north to Tacoma. It was made more efficient by a new crossing of the Cascades that brought it to Seattle on June 1, 1887.[2]

The Canadian Pacific Railway was completed through British Columbia in November 1885 and opened for regular traffic the following summer. William Cornelius Van Horne, tough, aggressive and competent, was manager and then became president of the new railway. Its initial terminal of Port Moody on Burrard Inlet was only a temporary end of track before the railway was extended west to the site of what soon became the booming port city of Vancouver. Next came the Great Northern Railway of James Jerome Hill, built across the plains and mountains of Montana, northern Idaho and Washington to Everett and Seattle on the shore of Puget Sound. Hill's line opened for through traffic in 1893. All three companies, the NP, GN and CP, were desperately in need of traffic to support the huge expenses of construction and to equip the lines for operation. And all three lines passed through hundreds of miles of lands largely unsettled and in many areas fairly recently depopulated of aboriginal peoples following epidemics, war and enforced movements to reserves or reservations.[3]

These were remote and often inhospitable lands that initially offered little hope of supporting a railroad, and between the three transcontinentals were many miles of mountainous territory without effective means of communication, yet there were also many promising areas of mineralization, immense tracts of forestlands and areas of rich farming and ranching potential. It was a vast, resource-rich area that held promise of being one of the most favoured in the history of the development of the continent. To develop these lands, vast tracts of which were owned by the Northern Pacific and Canadian Pacific railways through land grants, they embarked on ambitious and extensive settlement schemes and promoted industrial expansion. The Great Northern, which was built without land grants, nonetheless promoted settlement and agriculture with considerable zeal, an enthusiasm exuded by its head, James Jerome Hill.

For this rich territory along the international border between Washington and British Columbia, three major centres of commerce developed on the coast. Seattle and Vancouver, with their growing ports and being the terminals of the major railways, were assured of controlling the trade over vast areas of the interior. To the east, Spokane, in central Washington, became the heart of a vast, productive region of farm, forest and mineral lands often called the Inland Empire, and its influence and dominance as the largest nearby city and financial centre extended into southern British Columbia. Between these three business poles, there was considerable competition for economic influence and it was intensified by the border that ran in a straight line across the mountains and rivers irrespective of topography, travel routes, the traditional territories of Native peoples or the dreams of those seeking their futures and fortunes.

As industrial activity and settlement boomed in the 1890s, financial and commercial interests on the coast, and particularly in Vancouver, worried over an increasing focus on Spokane as the centre for the trade of southern British Columbia, and resented central Canada's control over commerce and manufacturing, as well as the strong ties of the Canadian Pacific with those interests. Consequently, they began to lobby for a direct "Coast-to-Kootenay" railway that would realign the economic focus of the southern interior, and shift the business profits of the Kootenay, Slocan and Boundary districts to the coast.

The usual pattern of speculative development followed with applications to the provincial legislature for authority to build railways along with appeals for construction subsidies and land grants. Provincial charters were required for railways built within British Columbia, and with appropriate influence and argument a proposal might also be given a federal charter with possible further grants of funds or lands. Far more charters were granted than the country could ever have supported as operational railways. Nonetheless, with suf-

The first CPR branch line built south of its main line through the Rockies was the short, but highly functional Columbia & Kootenay which followed the Kootenay River between its confluence with the Columbia and Kootenay Lake at Nelson. Its purpose was to bypass the falls and rapids of the Kootenay River and provide a connection between the steamer services operated south from Revelstoke down the Columbia and on Kootenay Lake. Completed in 1891, its opening complemented the construction of the steamer *Lytton* on the Columbia the year before. This scene, looking east, is just west of Nelson where the tracks crossed the Kootenay River over a long series of wooden Howe truss spans. Later, these were replaced with more modern bridges. The little 4-4-0 on the train is from the Canadian Pacific's stable of early locomotives that powered the railway's trains in its first years, but were soon inadequate for main line traffic. They were rapidly replaced by larger and more powerful locomotives and relegated to branch line duties.

–Canadian Pacific Archives, 4622.

CANADIAN PACIFIC RAILWAY KOOTENAY DISTRICT BRITISH COLUMBIA

ficient subsidies, a good economic climate and booming resource development, a charter could become a valuable commodity. Several charters became part of the story of Great Northern expansion in the 1890s and, initially, they centred on Vancouver.

As the Great Northern was being built across the northern states to Puget Sound, James J. Hill rightly concluded that he needed terminals on the coast and access to the major sources of traffic. Although he eventually settled on Seattle, with its fine port on Elliott Bay, as his major terminal he envisioned a north–south line connecting Vancouver, Seattle and Portland, as "the head of a rake" to feed traffic to the Great Northern main line and provide effective competition for the Canadian Pacific, Northern Pacific and Union Pacific which all reached the area.[4] Hill explored his options long before the rails were completed through the Cascades, and in the early 1890s began securing his connections. The northern extension of Hill's rake was a line to Vancouver, or at least New Westminster. Parts of this railway were envisioned and developed by Nelson Bennett, a railroad engineer of considerable talent, who along with business associates organized the Fairhaven & Southern and also secured the charter for the New Westminster Southern, intended to connect New Westminster with lines to the south. The backers of the New Westminster Southern included two prominent and influential British Columbia businessmen, Alexander Ewan and John Hen-

Canadian Pacific Fought Charters for Railways from the South...

15th. March, 1898.
To Hon. A.G. Blair,
Minister of Railways and Canals,
Ottawa
Dear Mr. Blair:

...this railway [the Kettle River Railway] crosses the International Boundary several times. This is done because the road can be built more cheaply by following the waterways. In this part of the country the passes all lie across the International Boundary, parts being in the United States and parts in Canada. For instance, a large reduction could be made in the cost of the road which we propose to build between Robson and Boundary Creek if we could follow the Kettle River through Washington instead of climbing over the mountain range by way of Brown and Eholt Creeks as in the one case we would have a broad valley to follow with no grade steeper than one percent, while remaining North of the International Boundary we have a summit 1255 feet [383 m] high to cross... If he [Corbin] is permitted to cross and re-cross the Boundary it will reduce the cost per mile... to less than one half.... Mr. Corbin can dip into Canada and... take such business as there may be [in

Thomas Shaughnessy's letter continues on the next page...

dry. Ewan was a wealthy industrialist with substantial investments in the west coast salmon canning trade and sawmilling and who had extensive land holdings in the Fraser Valley. Hendry was a successful lumberman with major foreshore holdings near Vancouver, and he tried to expand his interests and influences on a large scale, promoting his railway charters and related interests with, it seemed, any company that might be tempted to invest, which included for a time the Great Northern.[5] He also controlled the charter of the Vancouver, Westminster & Yukon Railway, which the Great Northern eventually used to gain its important access to Vancouver from New Westminster. Eventually, he fell out with Louis Hill, the son of James J. Hill, over his charter mongering and politicking, but he profited well from his association with the Great Northern, apparently receiving substantial payments for his services despite his very divided and expensive loyalties.[6]

The GN acquired control of these north–south railway interests that provided the legal basis for most of their line along Puget Sound to Vancouver in 1888–1889, and extended the Fairhaven & Southern from Bellingham north to the border, and built the New Westminster Southern from the south shore of the Fraser River to join the line to Bellingham at the border. This route was in operation by 1891 and its incursion into the lower Fraser Valley generated some concern from Canadian Pacific. All that remained was an efficient crossing of the Fraser River and an entry into Vancouver itself.[7] The next piece of the puzzle began to take shape with the granting of a charter on May 8, 1897, for the Vancouver, Victoria & Eastern Railway & Navigation Company, usually called simply the VV&E, and it was to become a key to the story of railway development across southern British Columbia and northern Washington. Legislation was passed the next year granting a federal charter for the railway.[8] The incorporators of the VV&E were William Templeton, who was Vancouver's mayor in 1897, William L. Nicol of Vancouver, George Lawson Milne, a prominent doctor and later the Dominion Medical Inspector of Immigration at Victoria, John T. Bethune of Victoria, and Alexander Ewan of New Westminster, who had been involved with the schemes already noted. By this time Ewan and Hendry had developed, with Great Northern backing, the Kaslo & Slocan Railway in the West Kootenay, so their association with James J. Hill was well established.[9]

The projected route for the VV&E, as described in the charter legislation, was from Vancouver, at "some point on Burrard Inlet or English Bay," to New Westminster, eastward through the Fraser Valley and the "southern part of British Columbia by the most feasible route, to the City of Rossland." Along the way, it was to pass through Chilliwack in the Fraser Valley, the Hope Mountains (the Cascade Range) and the Boundary mining district. The railway was required to build south of the main line of the Canadian Pacific but was also given permission to build a branch line to Kamloops, and south to "some point on the coast line between the International Boundary line and the Fraser River." The company was also empowered to operate steamships on the coast and on the interior lakes and rivers. The province offered a cash subsidy, sometimes termed a "bonus," of $4,000 a mile (about $2,500 per km) once the trackage was completed. A further condition of the subsidy was that the VV&E operate a steamer service between the mainland and Vancouver Island.[10]

Later in 1897, whatever the interest of James J. Hill was at that time, the VV&E's provincial charter came under the control of businessmen and contractors William Mackenzie and Donald Mann. For some years Mackenzie and Mann and other eastern businessmen had been investing in the mining developments of southern B.C., so their railway interests were well integrated into their other concerns. However, the provincial subsidy for the VV&E was contingent on the company obtaining a federal charter and this had not yet been approved. Mackenzie and Mann were also hoping to build a rail line through the Crowsnest Pass, just north of the international boundary in the Rockies, and the VV&E would have provided the connection through to the coast. However, the Laurier government in Ottawa did not favour the Mackenzie and Mann proposal because it lacked eastern connections that would have permitted highly desired lower freight rates to the Lakehead in Ontario. Instead, the CPR began construction through the Crow. A change in government in British Columbia followed and Charles Semlin, the new premier, postponed the VV&E project in August 1898, claiming that the contract was void because no federal support had been gained although a federal charter had in the end been granted.[11] Survey crews had already been in the field and Mackenzie and Mann considered court action. However, political instability continued in British Columbia. A new government under James Dunsmuir, coal magnate and owner, with others, of the Esquimalt & Nanaimo Railway on Vancouver Island, was more supportive and the VV&E scheme was kept alive.

In the end, Mackenzie and Mann were not prepared to build the railway on their own without subsidy and they were unsuccessful in obtaining one. In March 1901 they came to an arrangement with James J. Hill, who had in 1898 already acquired the railways built by Daniel Chase Corbin to Nelson and Rossland, and who saw the VV&E charter as a means of extending his railway interests in southern British Columbia, and potentially across the Canadian Prairies. With the VV&E charter in hand, Hill had the means to begin construction into an area that he recognized could offer considerable traffic from the growing mining districts close to the border, attract some of this revenue to his main line through Washington, Idaho and Montana, and at the same time pressure the CPR over what he saw as its incursions into the Midwestern states through its Soo Line subsidiary.[12] "The root of the conflict," explains Ralph W. Hidy and his collaborators in their extensive history of the Great Northern, "was Hill's obsession with offsetting the CPR's advantage—through its Soo Line—in rate-making from Midwestern points to the Pacific Coast, his determination to set up GN affiliates in Canada comparable to CPR's Soo in the United States, and his eagerness to protect and extend the northern frontiers of his empire."[13]

By the late 1890s William C. Van Horne's time as president of the Canadian Pacific was nearing an end. He had differences with George Stephen, former CPR president and important financier, and also an investor in Hill's enterprises. Van Horne acquired interests in the trackage of the Duluth & Winnipeg with its lucrative ore reserves, a railway highly desired by Hill, and it was a point of some tension. The CPR eventually relented and Hill acquired the road, but Van Horne felt it time to move on and resigned as president in 1899, leaving the railway in the able hands of Thomas Shaughnessy who assumed the presidency. Van Horne then became the board chairman. In practice, from that time on the Boundary lines were a Shaughnessy enterprise, not one that would be closely connected to Van Horne. Van Horne turned his energies elsewhere including building railways in Cuba, and in central Cuba, near the railway, a century later you can still find a sign for "Von Horne" Ave.[14]

Relations between the Great Northern and the Canadian Pacific had been strained by the late 1890s and the early 1900s as the two giant enterprises cautiously wrestled for control of key traffic areas, lobbied for political leverage and used incursions into each other's perceived territory in a point-counterpoint exercise that headed them for very direct competition in the Boundary and Similkameen districts of B.C.

Much has been made of the rivalry between the two lines and between Hill and Van Horne in particular, but it was a complex relationship, not given to simple assertions of hostility or rivalry. And both parties made mistakes, and misjudged situations, acted on incomplete or incorrect information, or overreacted. However much business was indeed business, and they played a hard fought game, it did not preclude the Hills being invited to vacation with the Van Hornes. Officials from either railway, who remained impeccably civil and were often on friendly terms, could sound pious and innocent, aggressive or conciliatory, businesslike, patriotic, wounded or temperate. The companies were a good match; but perhaps not a match made in heaven, and they probably deserved each other.

John Hendry was president of the VV&E under Hill's financing, but he promoted his own interests and fell out with Louis Hill before the arrangement ended in 1907.
–Robert Turner Collection.

Shaughnessy continues...

the Boundary District] and carry it out of the country with a railway which will cost only half as much to build as a line serving all the territory in Canada between Robson and Boundary Creek.

I can quite understand that mine owners and others in the Boundary Creek District anxious for a railway communication would be willing to have the Corbin line rather than none... but with the assurance that such communication will be obtained this year by a line entirely within Canada they would have no justification for favouring the Corbin scheme; indeed, I feel confident that not one of them who is a Canadian would favour it under such circumstances.

Yours very truly,
T. G. Shaughnessy,
Vice-President.[15]

Phoenix and the Mountain of Copper

When the Railway Came to Town...

"Amid the shrieking of whistles and cheers of the assembled miners and other citizens of Phoenix and with the ore bunkers gaily decorated with flags the last piece of steel was laid to the old Ironsides ore bins at noon today. Immediately after the last spike was driven the deep basso-profundo whistle of the Ironsides shaft house heralded the welcome news quickly followed by the Knob Hill and in turn taken up by the War Eagle, Gold Drop, Golden Crown and other mines. The terrible earsplitting din was not lessened by numerous explosions of dynamite and C.P.R. locomotive whistles."

—*The Greenwood Times*, May 23, 1900.

The Mines and Smelters Were Intensely Political...

"The great importance to every section of Canada of having our ores smelted in Canada, and large smelting towns built up within our borders, cannot be over-estimated. We, of course, have a selfish interest in doing everything we can to forward and safeguard this policy, but our interests are exactly in line with those of every merchant, manufacturer, and producer in Canada. Such legislation as that proposed by the Kettle River Railway bill should receive a sledge hammer blow from every commercial community in the Country."

—Thomas Shaughnessy to W. D. Matthews, Board of Trade, Toronto, Ont., March 23, 1898.

Phoenix, on the top of the mountain ridge between Grand Forks and Greenwood, at 4,603 feet in elevation at the Canadian Pacific station and 4,361 feet at the rival Great Northern depot (1403 m and 1329 m respectively), was the highest city in Canada in the early 1900s. It was the focus for a mining boom that saw it prosper, fade away and eventually be devoured by an open-pit mine that a generation later consumed the townsite. The Phoenix mining boom surpassed anything the southern interior had seen before. Carloadings from Phoenix outranked those from all but two other B.C. cities, and the two railroads were engaged in an around-the-clock struggle to transport 5,000 tons[1] of copper ore each day to Grand Forks and one of the largest smelters in the British Empire, which was located only 24 miles (38.5 km) away but 3,000 feet (over 900 m) below.

This mining boom in the Boundary District had its beginnings in a much earlier mineral discovery. On the hot afternoon of July 15, 1891, prospectors Henry White and Matthew Hotter laboured up the steep, wooded northwest slope of Knob Hill in the rolling mountains high above Boundary Creek, a tributary of the Kettle River, not far north of the British Columbia–Washington border. They travelled over two miles (over 3 km) from the frontier mining camp of Skylark, located on the hill southeast of what was to become the city of Greenwood. High above the pretty valley to the west, they made their discovery on the south slope of a basin at the head of Twin Creek. There White staked what was to become the famous Knob Hill mine and his partner staked the Old Ironsides claim immediately to the north. Although the claims were not filed until the

The Boundary Creek, Most Important of all...

To the Editor of the *Globe,*

Sir,

I am astonished beyond measure at the editorial in the *Globe* of yesterday on the Kettle River Railway. The responsible Editor must have been abroad, for the fairness and the good business judgement and the patriotic sentiment which usually characterize the *Globe* seem to me to be all wanting.

If the Boundary Creek district is as rich in precious metals as is claimed by those who know it best, the tapping of that district by the American lines will result in a loss to Canada of *one hundred million dollars* within the next ten years, but were it only one tenth of that it would be a sin against the country to let it go.

Yours very truly,

W. C. Van Horne
19th March, 1898.

Photo, facing page...

An ore train from Phoenix, with a mid-train helper, rumbles across the timber bridge over the Granby River just downstream from the dam that formed Smelter Lake. The wooden flume to the powerhouse is in the foreground.

–Vancouver Public Library, 2372.

Jay P. Graves was one of the most prominent figures in the mining development of Phoenix and the creation of the Granby Company which operated the huge smelter at Grand Forks. The Spokane businessman is shown above left, probably at Grand Forks, with two companions who are unfortunately unidentified. Graves went on to a prominent career in Spokane involving land development, interurban railways and other business enterprises.
–Boundary Museum, 986.253.001.09.3

The Granby Company...

"The Granby company was floated in Montreal.... Mr. J. P. Graves has in the floatation of the two mining companies and the new smelter demonstrated that he is one of the cleverest mining operators in the west."
–*The Boundary Creek Times*, December 17, 1898.

following Tuesday, a flurry of activity by Greenwood-area prospectors ensued. Several other claims were located including the Stemwinder on the north side of Twin Creek, staked on July 25 by J. Attwood and J. Scholefield. Soon after, Edmond Lefebvre staked a fractional claim between it and the Old Ironsides. He called this the Silver King, but like several other claims, it was allowed to lapse and was eventually relocated by Robert Dengler, who named it the Phoenix, and the immediate vicinity soon took on the name Phoenix Hill. The city of Phoenix was to develop over and around this claim although at the time the area was known as the Greenwood Camp, the easiest access being via Greenwood in the narrow valley below to the west.

Many more claims were staked that summer. On July 31, Joseph M. Taylor and Steve Mangott staked the Brooklyn adjacent to the Stemwinder on the northwest side. A few days later Bob Dengler and Willis Douglas located the North Star on the south slope of Twin Creek valley below the Old Ironsides. As happened so often, a promising first claim led to a number of others being taken up in the immediate vicinity. Some were speculative and allowed to lapse but most were soon taken up by other speculators and investors, or were relocated, as was the Brooklyn in 1894 by John Meyer and George W. Rumberger.

When earlier claims in the Deadwood Camp, a short distance northwest of Greenwood, where the Mother Lode, Crown Silver and Sunset claims had been staked in 1891, were added to those of the Phoenix area, it became obvious that the Boundary District was on the threshold of a massive expansion, even if only a few of these early prospects proved to be profitable.

Continued surface work showed that the Phoenix Hill claims contained enormous ore bodies, but the grade of ore was so low in copper values that large-scale development was premature. The lack of rail connections and the distance from smelters would hinder development for some time to come. By the end of 1894, 370 claims had been staked in the Boundary region, but the economic realities of the low-grade ore, the remote site and the impact of the severe economic depression of the early and mid-1890s dramatically restrained investment in these properties and only 127 claims were maintained. Nonetheless, more prospectors and miners came into the district and the staking of new claims escalated. In 1895, 771 new locations were registered and the next year 1,279 more were staked.

Henry White and Matthew Hotter continued to develop their Knob Hill and Old Ironsides properties and in 1895 they sold a part interest to Jay Paul Graves of Spokane. Graves was an enterprising and ambitious man, then just 35 years old, who was to have a profound influence in the Boundary District and the Inland Empire around Spokane. Initially a man with little financial backing, he used charm, skill and sheer determination to build a substantial personal fortune.[2] Graves and his new partners formed the Old Ironsides Mining Company, and in 1896 they organized the Knob Hill Gold Mining Company. Financial backing was not easy to attract, especially among westerners who were finding the recently discovered Rossland mines with their better transportation routes and smelters at nearby Trail Creek, B.C., or Northport, Washington, to be more promising investments.

A breakthrough came with the discovery that the Phoenix ores contained enough lime to make them practically self-fluxing when smelted and, as a result, much easier and more economical to process. Graves, in association with Aubrey L. White of Montreal, managed to obtain the backing of Stephen C. H. Miner, president of the Granby Rubber Company of Granby, Quebec, a director of the Eastern Townships Bank, and a man with a promoter's confidence and flair. Using his charm and influence he soon obtained financing for the development of their mines in the Boundary. In 1896 Miner and his partners from Quebec formed the Granby Consolidated Mining & Smelting Company and joined Graves and his associates to form the Miner-Graves syndicate to develop all of their ore bodies under one company. Miner also made sure that the booming mining towns that flourished near the new mining properties were well served by branches of the Eastern Townships Bank.[3]

By this time, the main shaft of the Old Ironsides had been put down a mere 100 feet (30 m) and the Knob Hill tunnel was not much more advanced. To speed production, a 60-horsepower steam plant was hauled in over the rough wagon road from Marcus, Washington, the closest railroad point to the district. Marcus, 60 rugged miles (nearly 100 km) distant, was located on the Spokane Falls & Northern Railway (SF&N), which connected through to Spokane.

By the spring of 1897 the steam plant was in full operation, and powered two drills that considerably speeded up development work. William Yolen Williams, a soft-spoken, experienced miner from Wales who had been at Rossland, was placed in charge as general superintendent. At the Old Ironsides the first half of a ten-drill air compressor was installed, followed in a few months by the second half, which also supplied the Knob Hill drills. By the end of 1898 there were over 2,000 tons of ore on the dump at the Old Ironsides Mine, and exploratory work showed indications of huge tonnages available.

The key to bringing the mines into profitable production was rail transportation to move the immense tonnages to smelters, and then to markets. Moreover, with low-grade ores, it was best to have concentrators and smelters as close as possible to the ore supplies to minimize the expensive haulage of waste materials. Geographic barriers and financial connections focused the interests of the Boundary District on Spokane, the centre of Washington's Inland Empire. Railroads fanned out in all directions from Spokane on the main line of the Great Northern Railway completed in 1893, and since 1890 the main artery northward had been the Spokane Falls & Northern, owned by Daniel Chase Corbin of Spokane.[4] In 1893, Corbin extended his railway northeastward to Nelson, B.C., calling his Canadian section the Nelson & Fort Sheppard (the initials of the N&FS being the reverse of the SF&N), only to have the mining interest shift westward to Rossland. Undaunted, Corbin completed the Columbia & Red Mountain Railway from the Spokane Falls & Northern at Northport, on the Columbia River, north to the Canadian border. From there the line was built as the Red Mountain Railway, which snaked up to Rossland around 24-degree horseshoe loops, to reach the rich mines in the mountains high above the river. At Rossland Corbin came into direct competition with Augustus Heinze and his Trail Creek Tramway that was built to take the Rossland ores to a new smelter built by Heinze on the west bank of the Columbia at Trail.[5] However as the Rossland–Trail mines and smelter developed there was ample traffic for Corbin's lines and much traffic flowed south.

Initially, Canadian Pacific's presence in the southern mining districts was limited and roundabout. Steamers operated down the Columbia from Revelstoke to Robson and Trail. From Robson the CPR's Columbia & Kootenay Railway ran to Nelson. However the CPR recognized its vulnerability and soon took steps to improve its position.

After building to Rossland, Corbin turned his attention westward to the Boundary District in B.C., and to Republic, about 35 miles (55 km) to the south of Grand Forks, in the Colville Indian Reservation. The reservation, established in 1872 by executive order of President Ulysses S. Grant despite no treaty with the Colville people, included large areas of potentially rich mineral deposits and timber, too rich to avoid the avarice of capitalists, miners and speculators and their demands for opening up the area. The northern half of the reservation was opened to mining exploration on February 21, 1896, and in anticipation, prospectors had already located some claims well in advance. The discovery of the Black Tail claim just four days after the opening, followed soon after by the Lone Pine claim, set off a wave of frantic staking in Eureka Gulch near Republic, and by 1898 a number of promising copper-gold properties were being developed in the area and needed only a railroad to transport their rich ores to smelters. In 1898, the southern half of the Colville Reservation was also opened for mining development, and in the rush that followed (and as before, unofficially often preceded) the opening, 5,000 claims were staked

Shaughnessy Protests to Prime Minister Laurier...

31st. March, 1898.

PERSONAL

Dear Sir Wilfrid,

There can be no doubt that the course pursued by the Minister of Railways when the Kettle River Railway Bill was before the Railway Committee was accepted by many of the Government supporters as conclusive evidence that the Government favoured the passage of the Bill.... I cannot believe that the Government as a whole... calculated to seriously damage the business interests of British Columbia... The purpose of the line is... to enable Mr. Corbin to take the ores from the mines [in the Boundary] to his smelter at Northport for treatment.... If the smelters at Northport or Spokane or Helena are to be given free access to our choicest ores the smelting business will be lost to Canada. We feel it a duty to the Company and to the country in whose welfare we have such a very large stake to fight this Bill to the last stage.

T. G. Shaughnessy,
Vice-President

Charter Mongers for Paper Railways...

The VV&E charter mongers received the treatment they deserved from the Provincial legislature last week. They used their influence in and out of Parliament to keep Heinze from building into the district...[and] influence to stop Corbin from building. They hawked the rights accorded them by the legislature and bandied their charter as an asset they could dispose of to the highest bidder. The government cancelled the subsidy for the road from Point Roberts to the Boundary Creek District. The act repealing the subsidy was passed 20-13."

–*The Boundary Creek Times*, March 1, 1899.

Sale of the VV&E...

"It is understood that the V.V.&E. Ry. & N. Co. is now entirely controlled by the Great Northern Ry... Mackenzie, Mann & Co. previously held a considerable interest in the charter."

–*The Railway and Shipping World*, March 1903, p. 108.

in the first week. Eroding the Colville Indian Reservation even further, two years later the northern half was opened for homesteaders and timber cutting and some years later the southern half also was opened for these purposes.[6] In the 1890s, this isolated area could best be reached economically by a railway built from Marcus on the SF&N. The most feasible route followed the Kettle River northward, crossed the international border, then turned westward for about 15 miles (25 km) to a point south of Grand Forks before re-entering Washington to follow the Kettle River and Curlew Creek south through to Republic.

On March 19, 1898, the Kettle River Valley Railway Company was incorporated in Washington State by Daniel Chase Corbin's son, Austin II, his chief engineer Edward J. Roberts and Jay H. Adams, a Spokane attorney. But Corbin underestimated the power of the Canadian Pacific when he tried to gain a charter in Canada for the sections north of the border. Although no subsidies were requested, and petitions were presented backing the proposal from the citizens of Greenwood and Grand Forks, the submission for the line to Andrew George Blair, federal Minister of Railways, still met with determined opposition. In Ottawa, the proposal passed the Railway Committee by a vote of 54–48 over the heavy opposition of the CPR, but the scheme was doomed. The CPR's William C. Van Horne claimed that the Boundary District was the richest mining area known in the world and that, if Corbin were allowed to build this railway, the line would take a fortune out of Canada. Opposition also came from the British Columbia provincial government, which generally opposed any charters that might divert traffic to the United States. In consequence, on April 15, 1898, the KRVR charter was defeated 64–44 in the House of Commons, with the CPR allegedly providing special trains to carry members sympathetic to their view from their Easter vacations for the vote. Corbin was said to have remarked, "The Canadian Pacific used all its resources to defeat me."[7]

Corbin's proposed line was an inexpensive-to-construct, water-level route into the Grand Forks valley, with no grades steeper than 0.6 percent and low operating costs. It would have been cheaper to build than an all-Canadian route, and cheaper to operate, a point the CPR hastened to emphasize in its opposition. It might have made a great deal of sense economically, even though it failed politically. The Canadian Pacific was left with a clear field into the Boundary District at a time when the company was already expanding into the Kootenays on several fronts, making up for its previously poor rail and water connections to the mining districts. At the time of Corbin's proposal, the CPR had a lot at stake and a substantial investment in new construction that had yet to see any returns. By the summer of 1898, construction was nearing completion on a new route through the Crowsnest Pass, rich in coal deposits, from southern Alberta to the southern end of Kootenay Lake. In connection with steamer services on Kootenay Lake and the short, but important, rail connection built between Nelson and Robson on the Columbia, the CPR had extended its influence across the southeastern mining districts of the province. In 1896, the CPR had purchased the Columbia & Kootenay Steam Navigation Company that operated the largest fleet of steam vessels in the region and included routes on the Columbia River between the CPR main line and northern Washington as well as on Kootenay Lake. Expanded steamer services and branch line construction south from Revelstoke to Arrowhead, from Nakusp on the Columbia to Slocan Lake, and from the southern end of Slocan Lake to Slocan Junction on the Nelson–Robson line greatly increased the CPR's access to the mining districts of southern British Columbia and helped prevent further incursions of American lines from the south.[8]

Although the CPR's rail and steamer operations were modern and well equipped, and provided frequent service, the routes were still rather roundabout. A trip from Nelson to Vancouver took several days and required journeys by rail, steamer and rail again, whereas the Corbin lines, along with the Great Northern, offered direct rail services to Seattle, with good connections to Vancouver, and Spokane was just hours away.

Photographer Richard Trueman, a master of his art who worked at Revelstoke, Kaslo and other parts of the Kootenays, as well as in Vancouver, recorded this scene looking across the Columbia at Robson toward the steamer dock and the railway facilities. Before the completion of the railway bridge across the river south of this point in 1902 (shown on page 52), sternwheelers shuttled back and forth across the river with barges carrying passengers, freight and railcars to the railway on the west bank of the river. At the dock is the steamer *Lytton* near two passenger cars and the station. Steaming down the river is the little sternwheeler *Illecillewaet,* used as a tug, for icebreaking and as a freighter. At this time both were owned by the Columbia & Kootenay Steam Navigation Company which was purchased at the end of 1896 by the Canadian Pacific Railway to become the core of its British Columbia Lake & River Service.
–Richard H. Trueman, National Archives of Canada, PA-51390.

Robson, Robson West and Castlegar...

The locations and names used by Canadian Pacific for its stations near what is now Castlegar are confusing. Until its railway was built westward towards Grand Forks, Canadian Pacific's western end of track (officially of the Columbia & Kootenay Railway between Nelson and the Columbia, bypassing the rapids of the Kootenay River) was Robson on the east bank of the Columbia River a short distance north of the Columbia's confluence with the Kootenay River. This site is across from the present city of Castlegar. Robson was also a steamer landing where sternwheelers provided a connecting service between the Columbia & Kootenay Railway and the Canadian Pacific's branch line from Arrowhead to the main line at Revelstoke, (and earlier directly by steamer to Revelstoke). Across the river another steamer wharf was established called West Robson and later Robson West, 1.7 miles (2.7 km) north of Castlegar station. For a time the original Robson, probably in an attempt to avoid confusion, was called East Robson in Canadian Pacific timetables, but it eventually reverted back simply to Robson.

After the railway bridge over the Columbia was completed on March 20, 1902, Castlegar's station was called Castlegar Junction. The bridge crossed the Columbia south of West Robson, and Castlegar itself became the point where the tracks joined at a wye and diverged south to Trail and Rossland (the CPR's Rossland Subdivision) and westward to Grand Forks and Midway (although for the first miles the tracks followed the Columbia north before swinging westward). The short section of trackage along the east bank of the Columbia to Robson was removed soon after the bridge opened, but the site remained a steamer landing on the Arrow Lakes route, and in particular a stop of the well-remembered SS *Minto*, until the end of that service in the mid-1950s. By the 1920s, Castlegar's station name lost the postscript "Junction."[9]

The Chief Engineer of Construction...

"Mr. W. F. Tye of Trail, B.C., is appointed Chief Engineer of Construction of this Company's line between Robson and Midway. He will superintend and have charge of all matters connected with engineering and construction work."
Columbia & Western Railway
Office of the President,
Montreal
23rd June, 1898, File 359.
–Shaughnessy, Letterbook, 60.

To the Contractors...

June 8, 1898
Messrs Mann Foley & Larsen,
St. Paul, Minn.
Dear Sirs.
Referring to your contract to this Company for the construction of the line between Robson and Midway. The Company desires to complete this work as quickly as possible, and it is understood that you will push it with the utmost vigor. While the contract provides for completion... on or before December 31st. 1898, the Company will not, in view of what you say about the bridging, exact any penalty to complete the work within the specified time, provided that you exercise satisfactory diligence... the Company's manager to be the judge... but in no event will ... completion be later than May 31st. 1899.
Yours truly,
T. G. Shaughnessy

With an eye on Corbin's planned expansion into the Kettle River country, the CPR had its chief survey engineer, L. M. Rice, and a party of 18 men working east and west out of Greenwood covering a route mapped out the previous fall by John D. Sullivan for Augustus Heinze's Columbia & Western. Sullivan, just 35, already was a seasoned engineer who knew the southern mountains and mining districts of British Columbia very well. While he and his crew were in the field that winter, negotiations were in progress between the CPR and Heinze for the purchase of his railway and smelter holdings. At the same time, Eugene Sayre Topping, one of the founders of Trail, was in Greenwood seeking support for expansion of the Columbia & Western, but sentiment in the town was solidly behind Corbin and, as *The Boundary Creek Times* reported, Topping came to town with 75 signatures on his petition and left town with the same number.[10]

The next week the CPR purchased Heinze's interests, including the Trail smelter (the British Columbia Smelting & Refining Company) and the Columbia & Western Railway. The railway included recently completed standard gauge trackage from West Robson, on the western bank of the Columbia, to Trail and a three-foot-gauge railway or "tramway" from Trail Creek to Rossland. The CPR's recently acquired sternwheelers and barges shuttled back and forth across the river between the Columbia & Western and the terminus of the Columbia & Kootenay Railway, the CPR's connection eastward to Nelson. Steamers also ran up the Columbia to Arrowhead at the northern end of Upper Arrow Lake. From there a branch line, completed in 1896, bypassed a difficult section of river navigation and ran 27.5 miles (38.8 km) north to the main line of the Canadian Pacific at Revelstoke. In April, the Canadian government granted the Columbia & Western formal power to lease its line to the CPR in perpetuity.[11]

Acquisition of the Columbia & Western Railway, under whose charter Heinze had built his line to Rossland, gave the CPR the legal authority to build along the Columbia between Trail and Castlegar and west to the Boundary District. Moreover, it provided for a provincial land grant of 20,000 acres per mile (5030 ha per km) of completed railway. The land did not become subject to taxation for 10 years following the transfer of title, and station grounds, structures and other similar property were free from taxation for 10 years after the completion of the line to Penticton and to the west. In addition, the C&W was to receive a federal cash subsidy of $3,200 a mile (about $2,000 a km) for its line to the Boundary District.[12] The cash subsidy certainly did not cover all of the costs of construction that in the end averaged over $40,000 a mile ($25,000 a km), but it was a welcome contribution, as was the land grant, which demonstrated support from the federal and provincial governments for the project to shareholders and would-be investors in the railway and the mining district. Some years later, in a move to indirectly subsidize further railway expansion including the completion of the Kettle Valley Railway, the B.C. government purchased land originally granted to the railway. In 1912, under the sympathetic administration of Premier Richard McBride, 4,168,334 acres (1 686 864 ha) of still unsold Columbia & Western and British Columbia Southern lands were acquired by the province at 40 cents an acre.[13]

With Corbin's charter defeated, the CPR proceeded rapidly with its plans to build westward from the Columbia River at West Robson, 100 miles (160 km) to Midway, passing through Grand Forks and Greenwood in close proximity to the richest and most promising mining areas in the Boundary District. William Francis Tye, the CPR's chief engineer, and a long-time associate of Sullivan, was placed in charge of the project.[14] Like Sullivan, Tye was just 35 years old, in excellent physical condition, and a very capable engineer. He was a strapping six feet, four inches (193 cm) tall; it is not hard to picture him striding along the ridgetops on his surveys. Sullivan was appointed his assistant. Tye had supervised GN construction in the Cascades and the team had worked together on the Great Northern-backed Kaslo & Slocan Railway in the Slocan mining district of the West Kootenays. They immediately set to work preparing plans and estimates for the project, dividing the route into four 25-mile (40-km) sections, each in the charge of an engineer. F. M. Young took the first 25 miles west of Robson, George Farr set up camp at the summit of McRae Pass, while Oscar Englund made his headquarters at Cascade City on the Kettle River. L. M. Rice took the remaining section, making his headquarters at Greenwood where he could also prepare surveys for further extension of the railway from Midway to the Okanagan Valley (Okanogan in Washington).

A relatively easy route from West Robson to Midway could have been obtained by following the Columbia River downstream to its confluence with the Kettle River in Washington and then going upstream and back into Canada near Cascade City and on to Grand Forks. However, this route would have had considerable mileage in Washington and for political reasons, this was unacceptable. Instead, it was necessary to build the line over two mountain ranges. All but 14 miles (23 km) of the 100 miles (160 km) were located on heavy grades winding around high rock ledges. This necessitated much blasting and tunnelling plus the use of millions of board feet of timber to bridge the many side canyons that had to be crossed.[15] Even in the short, comparatively level section near Grand Forks, it was necessary to bridge the Kettle River in four places.

From an elevation of 1,422 feet (433 m) at West Robson, the route climbed to 3,984 feet (1215 m) at the summit of McRae Pass[16] with 2,557 feet (780 m) being gained in 25.6 miles (41.2 km) of the total 30 miles (48 km) of right-of-way. Having attained this elevation, the line descended to 1,559 feet (475 m) at the first Kettle River bridge east of Cascade City. In the 14 miles (22.5 km) from Grand Forks to the summit at Eholt to the west, the change in elevation was 1,351 feet (412 m) while beyond the summit to the west, the line descended 1,183 feet (360 m) in slightly less than 18 miles (29 km) following Eholt Creek and Boundary Creek valleys to Midway.

Bids for construction were opened at the end of May 1898 with Daniel Mann, the Foley Bros. and Peter Larson being the successful contractors. To oversee the operations they hired, perhaps rather ironically, Major John F. Stevens, the experienced and capable engineer who had discovered, at least as far as railway surveys were concerned, Marias Pass in the Rockies and Stevens Pass in the Cascades for James J. Hill's Great Northern. In addition, Stevens had surveyed the Spokane Falls & Northern for Daniel Corbin. Subcontracts were let immediately for the grading, tunnelling and bridgework. Bridge contractors for the whole line were the Porter Bros., Andrew, John and Richard, together with contractors McArthur & Roberts. Sections containing short tunnels went to the partnership of Foss & McDonald & Olaf Olsen, and the construction team of Winters, Parson & Boomer of Butte, Montana, was also represented. They had built the Columbia & Western from Trail to West Robson and would soon get the contract to rebuild the narrow gauge between Trail and Rossland to standard gauge. Others represented included Peter Larson's son-in-law, Patrick Welch, and the team of Burns & Jordan. The largest subcontract went to the McLean Bros., who were still working on the new British Columbia Southern Railway through the Crowsnest Pass to the east. They were to build a 3,000-foot (915-m) tunnel south of Bulldog Mountain to carry the line westward from

William F. Tye, in his office in 1904 at left, was in charge of CPR construction on the Columbia & Western between Robson and Midway. His chief assistant was John Sullivan, above, shown later in his distinguished career. The two men had worked together on other railway construction projects in southern British Columbia. When they worked on the Columbia & Western both men were in their 30s. Tye reported frequently to Vice-President Shaughnessy as construction proceeded. Some of Shaughnessy's instructions to Tye, very formal in the business tone of the day, are quoted in subsequent sidebars. A station on the line between Nelson and Kootenay Landing was named Tye when this rail link was finally completed in 1930.

–A. H. HARRIS PHOTO, MCCORD MUSEUM, NO 6488.N, ABOVE LEFT; ROBERT TURNER COLLECTION ABOVE.

Teamsters, like these on the road to Republic with their horse-drawn wagons in 1898, provided the remote Boundary mining communities with needed supplies and transported high-grade ore to the end of the railways. They also supplied the railway construction camps.

–Wilson Collection, Okanagan Archive Trust.

the valley of Pup Creek to the Dog Creek drainage. The tunnel was a huge and difficult project in such a remote location. The bore was to be 21 feet high by 16 feet wide (6.4 x 4.9 m) and an estimated 28,000 cubic yards (23 400 m^3) of rock—85,000 tons—had to be drilled, blasted and removed from the tunnel. In addition, they were to build a temporary route with switchbacks, all on 4 percent grades, to lift the line over the saddle of land above the tunnel while the tunnel was being driven. At least 300,000 ties were needed for the track.[17]

A wagon road to aid in construction was contracted to a man named Brown, who by mid-June had his crew grading the road from a point on the west shore of Lower Arrow Lake opposite the village of Deer Park about 15 miles (25 km) west of Robson. The route led up Pup Creek to intersect the railway right-of-way 1,200 feet (365 m) above the lake. By the end of the month, supplies were being forwarded up this road 25 miles (40 km) to the west. On the lakeshore at the mouth of Pup Creek, a log raft containing 370,000 board feet of timber was assembled to provide lumber for the 30 or more buildings under construction at the town of Brooklyn. Julius and Simon Peterson of Trail had already built a large hotel and the CPR and Dominion Express offices were in operation presided over by agent Peter Huckerley. Other construction camps were located at Shield's Landing, about eight miles (13 km) down the lake at Moberly Creek and at McCormick's Landing nearer West Robson.

Construction work on the climb from the Columbia River was exceedingly heavy. It was estimated that about 1.75 million cubic yards (1.46 million m^3) of rock had to be moved in the 15-mile (24-km) section between West Robson and Brooklyn. Peter Genelle of Nakusp, who had received a contract to supply 14,000,000 board feet of lumber for bridge construction, built a mill capable of cutting 60,000 board feet a day, two miles west of Robson. At the same time, McPherson & Stout established a 40,000-board-feet-a-day mill near the summit of the pass close to a number of locations where large bridges were required.

In July, the McLean Bros. completed their contract on the Crowsnest Pass line, moved their outfit to Brooklyn and commenced work on what at the time was to be the longest tunnel in B.C. Their initial equipment, including two large air compressors powered by three 80-horsepower steam boilers, two large air reservoirs, 14 drills and pipeline, cost $22,000 installed. This heavy machinery was delivered to the steamer dock at Brooklyn and hauled to the tunnel site. Manpower for the drilling was in short supply and remained so until completion of heavy construction on the Crowsnest branch of the CPR in October.[18] Despite the shortages of labour, about 90 men were available for grading the switchback route over the summit—a temporary line that, nonetheless, was to remain in service for a number of months after the opening of the line to Midway.

The general wage for workers on the grading was $2.00 a day and the cost of board was $5.00 a week. Tunnel workers got anywhere from $2.25 to $3.00 a day, depending on their experience. But even when skilled men were available, the work went slowly. For most of the time, water was in short supply and only one boiler could be kept under steam; consequently, just four rock drills could be operated at a time and work was delayed.

Progress on the wagon road was much better and by August the rutted wagon trail was 20 miles (32 km) beyond Brooklyn and being extended towards Christina Lake at a rate of about a quarter mile (400 m) a day. As the road advanced, a second construction town sprang to life at Gladstone (later renamed Coryell) in McRae

Shaughnessy's Instructions to Tye...

"In view of the character of the country, many branch lines necessary to reach mining camps and other sources of traffic might be rendered impracticable, or at any rate difficult of construction, if somebody else covered the ground first. Therefore, it is important that we should file under the Railway Act plans, profiles, and books of reference of any such lines in ample time to forestall any similar action upon the part of those who may succeed in securing charters.... The first plans to be filed should be those covering the best route up the Valley of the Kettle River and of lines necessary to protect us against encroachment of any railways that may cross the frontier between Christina Lake and Rock Creek Camp."

12th November 1898.

Continued on next page...

Pass, 24 miles (almost 40 km) from Brooklyn. This became the distribution point for the sections to the west. Between Gladstone and Cascade City at the foot of the lake, the roadbed lay on exposed ledges susceptible to snowslides and crossed numerous side canyons in all requiring 15 trestles in as many miles, while the total distance lay on a steady 2 percent grade. One of the most impressive trestles was a curved structure over Sutherland Creek, 10 miles (over 15 km) beyond Gladstone. It was 250 feet long and 70 feet high (76 x 21 m). Two other large trestles were required to cross Quinn and Porcupine creeks. The Porcupine trestle alone required over 1,000,000 board feet of timber and was 705 feet (215 m) long.

In mid-October 1898, William C. Van Horne, president of the CPR, paid a visit to Brooklyn and agreed with newsmen that the Midway branch was the hardest section of railway construction the CPR had yet encountered, and that the difficulties were much greater than anticipated. By that time the grading extended as far west as Greenwood and 5,000 men were employed on the project. The monthly payroll exceeded $200,000.

By early November, many of the grading contracts on the east end of the new railway had been completed and crews moved out to work elsewhere. Along with other labourers and tradesmen, many recently laid off on the Crowsnest line quickly found work on a new Great Northern branch line, the Bedlington & Nelson, being built from the Idaho border to the southern end of Kootenay Lake at Kuskonook. This line was controlled by the GN's Kootenay Railway & Navigation Company, owners of the Kaslo & Slocan Railway and several sternwheelers operating on Kootenay Lake.[19] Hill was aggressively building into British Columbia and posed a growing threat to the CPR's control of the rich mining districts along the international border. Adding to the effectiveness of Hill's position was his successful acquisition of Corbin's Spokane Falls & Northern Railway including the lines to Rossland and Nelson.[20] With the prospects of the Boundary and Republic mines looking better by the week, Hill and the Great Northern could not afford to overlook this potentially lucrative source of traffic as another area for expansion. Just like the CPR, the GN needed freight by the carload and passengers to sustain its long main line, which ran through many miles of still wild territory that produced little revenue.

Track-laying on the CPR line to Midway began in November 1898, the company having transferred an 80-ton Consolidation-type locomotive from the Selkirk section of the main line to supplement the C&W locomotives on the heavy grades. This new locomotive, CPR No. 403, was assigned to track-laying trains and, with the onset of heavy snow, the track had been put down as far as the bridge over Moberly Creek above Shield's Landing on the Columbia.

Grading continued along the lower sections, where snow levels were moderate, all through that winter, and in January 1899 Pat Welch set off a huge blast at Rattlesnake Bluff above Christina Lake. Three hundred kegs of black powder exploded in a mighty, reverberating detonation that shattered several thousand cubic yards of rock; windows were rattled and buildings shook in Cascade two miles away. Pat Welch was also the subcontractor for one of the toughest sections on the west end that included several miles of grade on a rock ledge high above the valley of the North Fork of the Kettle River. The construction camp of Niagara literally had to be cut out of the mountainside. A major obstacle was a 700-foot (215-m) tunnel that Foss & McDonald crews had been hand drilling and blasting. By March, they still had 300 feet (90 m) to go while Olaf Olsen, subcontractor, working on a 400-foot (120-m) tunnel east of Gladstone, had a mere 40 feet (12 m) left to complete.

The real bottleneck on the whole project was the Bulldog tunnel where only 400 feet (122 m) of the required 3,004 feet (915.5 m) had been completed by that time.[21] The McLeans continued to be handicapped by a lack of water, due to freezing conditions in the mountains, that severely limited the number of drills that could be operated. Track-laying was delayed by bridges that were still unfinished, but by April it had advanced to the big trestle over Pup Creek just east of the Bulldog tunnel. The Pup Creek Bridge provides a good example

Shaughnessy's instructions, continued....

"I want this to be the cleanest and most economical piece of railway work that we have ever done, the contractors receiving every penny that they are entitled to but not one penny more. No section or division engineer should have any latitude in this matter.... The monthly returns and everything else connected with the work up to this time have been most satisfactory. There is no reason why you should not make a record on this work that will establish you with us in just such a position as you desire to attain."

16TH DECEMBER 1898. FILE 106.

"I was rather horrified at your estimate of the cost of the line between Robson and Midway, namely $32,000 per mile exclusive of rails and fastenings, switches, frogs and so forth.... Are you personally taking precautions to check over the measurement and classification made by your division engineers, and are the cross sections being kept in such shape that a re-measurement can be made... when the work is completed, if it is considered desirable?"

26TH APRIL 1899. FILE 92.

All along the railway, section houses, water tanks and stations were needed for operational concerns and track maintenance. This photo shows the newly completed section house at Coryell on the route over Farron Hill between Cascade and Robson. The section houses provided accommodation for the section foreman and his family. Labourers were housed in bunkhouses. They were responsible for maintenance and keeping the tracks clear of rocks, slides or fallen trees and patrolled their sections on a daily basis.

–J. W. Heckman, Canadian Pacific Archives.

of the speed with which railway trestles were completed. It took exactly eight days for the bridge crew to erect and it contained 400,000 board feet of timber. By early May, the rails had been laid to the base of the switchbacks and to the east portal of Bulldog tunnel. Grading crews were busy removing the accumulated snow and mud from the cutting on the switchbacks.

As the warm weather of spring came to the southern mountains, the grading work was all but complete. Just a few miles along Eholt Creek and a few miles near Grand Forks remained to be let to contract. Grading on the spurs to the mining camps of Deadwood and Greenwood had been started in hopes of opening them as soon as track was in place on the main line. By midsummer, the tracks had been extended over the Bulldog switchbacks although much work remained on the high trestles from McRae Pass to Christina Lake. Of the 2,000,000 board feet of timber required for these bridges, only half was available in the Gladstone area; the rest had to be hauled in from West Robson.

In June the McLeans pulled out of their Bulldog tunnel contract after driving less than one third of the bore in 10 hard months of work including the long, snowy winter. Veteran tunnelman Olaf Olsen took over this contract and, using 120 men with 12 drills working constantly, and with warmer weather in the high country, managed to drive the tunnel ahead by an average of 10 feet (over 3 m) a day. By the first week of August more than 1,500 feet (457 m) had been completed.

As work progressed on the tunnel, construction trains used the switchbacks over the summit, bringing in supplies for the workers concentrating on the other major obstacle along the route, the huge 1,600-foot (488-m) timber bridge over the Kettle River just east of Cascade City, for which William Mackie was a subcontractor. Delayed by the delivery of ten carloads of stringers for the two 150-foot (45-m) centre spans, which had to be brought in from the coast via Revelstoke and the Columbia River route, the bridge was opened on August 12, 1899. Promptly at 10:27 a.m., gaily decorated locomotive No. 403, with H. N. Atkinson at the throttle and James H. Baxter firing, arrived at the west end of the long bridge over the Kettle to be acclaimed by the throngs of Cascadians as the first locomotive in the Boundary country. "The crowd at the bridge," noted *The Cascade Record*, "raised a subscription of $25.00 and shortly after J. A. Bertois drove up a wagon load of such refreshments as are most appreciated by labouring men on such occasions." The first through passengers arrived in Cascade that same day but had to travel the last few miles on a handcar and were put up in the Grand Central Hotel.[22]

The logistics of construction presented immense problems in a mountainous area like southern British Columbia. Not only did a small army of workers scattered along many miles of right-of-way have to be provided with food and lodging, but construction had to be coordinated in a systematic way. Usually, a pack trail or wagon road was built to construction camps that were established at key locations along the way. After the right-of-way was surveyed and all legal disputes resolved, the land had to be cleared before grading the roadbed could begin. At the same time, other construction crews were usually working on bridges, tunnels, cuts and fills, areas of heavy rock work and other points where there was likely to be a long delay. An incomplete major bridge or other obstacle could cause a serious delay in the entire process. Meanwhile, supplies of timber for bridges and ties, rails, spikes, track bolts, joint or angle bars and other materials had to be readily available for track construction.

Once the roadbed was graded, track-laying could begin. By the late 1800s, track-laying machines were in use. Specially constructed to operate with a locomotive and supply cars, they used conveyors to keep a steady stream of ties and rail moving up to the front of the train where the crews could put them in place. The rails were spiked down by the men as the train moved forward, all in a coordinated, continuously moving interplay of the construction workers, the track-laying machine and the train behind. Coming along next, other workers lined up the track and did any surfacing work needed. Next came the ballast trains and a final trimming and finishing of the tracks. Of course, the farther away the work was from sources of supply, the more complex were the problems of keeping everything running smoothly. Moreover, construction crews needed accommodations and food. Temporary camps were usually set up as close to the work sites as possible and then moved as the grading, bridge-building or track-laying progressed.

Reports from Cascade in August 1899 illustrate some of the complexity. "The second bridge, one west of Cascade, should be completed by Tuesday next and until then the tracklayers cannot proceed with the next 10-mile [16-km] stretch to third bridge. In the meantime, 1,500 feet [460 m] of sidetrack and several other spurs have been put down in the Cascade yards. Six locos are now in use by the contractor in handling freight and construction trains." After a week, "the tracklayers are now half way to Grand Forks. Ten days after crossing first bridge over the Kettle River, the track-laying machine with George Roberts at the levers and Conductor Haney in charge of the train crossed the second bridge and the same day put down 4,800 feet [1465 m] of line. Today

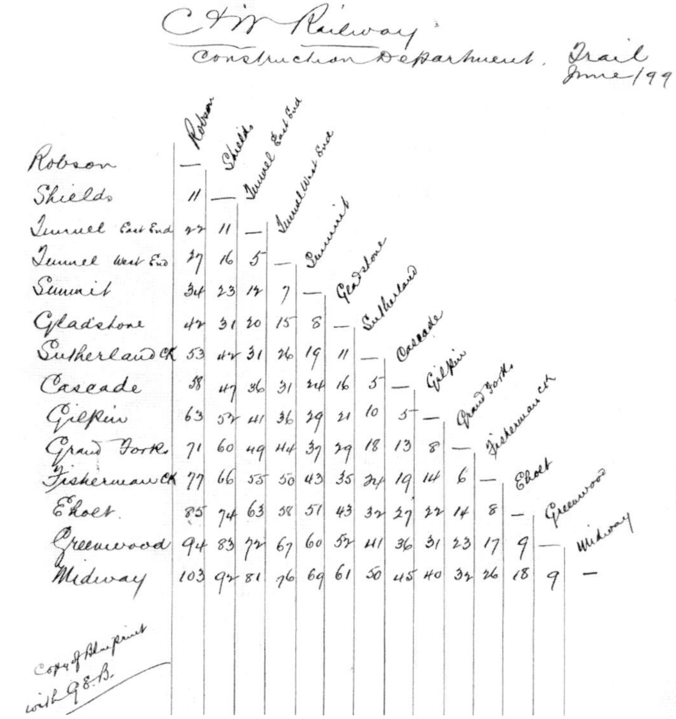
C&W Railway. Construction Department. Trail June/99

	Robson	Shields	Tunnel East End	Tunnel West End	Summit	Gladstone	Sutherland	Cascade	Gilpin	Grand Forks	Fisherman Ck	Eholt	Greenwood	Midway
Robson	—													
Shields	11	—												
Tunnel East End	22	11	—											
Tunnel West End	27	16	5	—										
Summit	34	23	12	7	—									
Gladstone	42	31	20	15	8	—								
Sutherland Ck	53	42	31	26	19	11	—							
Cascade	58	47	36	31	24	16	5	—						
Gilpin	63	52	41	36	29	21	10	5	—					
Grand Forks	71	60	49	44	37	29	18	13	8	—				
Fisherman Ck	77	66	55	50	43	35	24	19	14	6	—			
Eholt	85	74	63	58	51	43	32	27	22	14	8	—		
Greenwood	94	83	72	67	60	52	41	36	31	23	17	9	—	
Midway	103	92	81	76	69	61	50	45	40	32	26	18	9	—

Copy of Blueprint with G.E.B.

they finished the sidetrack at Cascade, put in a switch at Gilpin's [about halfway between Cascade and Grand Forks] so that boarding cars could be taken out to near Edwards ferry and put on a temporary spur. Engine 403 was taken to Trail shops on Tuesday for repairs and No. 143, an eight-wheeler, is being used in pushing the track-laying machine this week. From second crossing 56 lb. steel is being used instead of 73 lb. used on the hill. …8,200 feet [2500 m] laid yesterday."[23] Rail was measured in pounds per yard and heavier rail made it possible to use larger locomotives, or at least locomotives with heavier axle loadings, and larger freight cars.

Steadily, the tracks were approaching the community of Grand Forks, and once the flat valley lands were reached just east of the town centre, the residents could hear the whistles of the construction trains as the tracks neared town. Grand Forks, originally called Grand Prairie, was situated in a fertile, beautiful valley at the confluence of the mainstream of the Kettle River, flowing from the west and north, and the North Fork of the Kettle

At left is a table of distances along the C&W's during the building of the railway from. It includes "Tunnel East End" and "Tunnel West End," but the distances reflect the switchbacks. The C&Ws covered bridge, above, over the Kettle River west of Cascade was an enclosed structure.

–Canadian Pacific Archives, left; Boundary Museum, 987.222.48, above.

Passenger Service...

"Think best that you should take the line over to Grand Forks soon as possible, and operate it for account of company until road completed to Midway. The only service that would be required until the line is completed to Midway would be a sort of mixed train run as frequently as the traffic may require, and you should be able to get good stiff rates."

–T. G. Shaughnessy to W. Tye 10th August 1899. File 433.

Crossing the Kettle River at Cascade...

Intricate trestlework is still under construction on the approaches to the centre spans of the huge timber bridge over the Kettle River at Cascade in this photo from late spring, 1899. This same scene was photographed in 1967 by Dave Wilkie and is shown on page 8. At right, falsework is in place for the large deck truss bridge that carried the railway over the river. Light trestles were constructed and the more substantial truss spans were erected round the original timbers. The view shows the centre spans under construction. Once they were place, the temporary trestles were removed.
–Cominco Collection, left; Trail Museum & Archives, right, with thanks to Brian Wilson.

(now called the Granby River), draining from the vast forestlands that lay to the north. The Kettle meandered through the centre of the valley where, in the 1890s, a town and agricultural community was growing rapidly. Discoveries of copper and other minerals in the area gave promise of Grand Forks becoming a mining centre and the fertile soils encouraged farming a wide variety of crops. But like most frontier communities, its prosperity depended in large measure on having reliable and cheap transportation links to markets and suppliers.

West of Cascade, in the original survey, one more crossing of the Kettle River was required to reach Grand Forks. But local politics and land speculation led to an unexpected change in plans. Grand Forks civic officials objected to having CPR trains crossing their streets, and a new plan was drawn up hastily and the line was rerouted south of town at the considerable extra expense of two additional bridges over the Kettle River. The CPR was accustomed to local manoeuvrings to either attract the railway or to force excessively high prices in land deals. Some of the dealings, and foolishness, become clearer from letters between Grand Forks solicitor Richard Armstrong and T. P. Coffee of Toronto, who was a director of the Grand Forks Land Company.

"I had a long interview yesterday with the Chief engineer of the C.P.R. and their solicitor who came here to try to first jolly us into letting them cross our streets and failing that to bulldoze us into crossing them, but it did not work. I told them that I... would advise the city to arrest any employee of the C.P.R. or any other person who tampered with our streets or trespassed and if they interfered that I would advise our city to swear in special police and shoot anyone that would interfere with our streets.... They have gone... to try and get Mr. Shaughnessy to consent to build a depot and station in Grand Forks and I have no doubt as to the result as they have about as big a chance of getting away with us as a calf has with a butcher."[24]

The next day, the CPR's decision became clear and Armstrong wrote, "...the C.P.R. are going to abandon the line that they have graded and the bridge that they have partially completed and make a circuit around the town entirely. ...They will not be able to locate their depot one yard farther away from us than they have

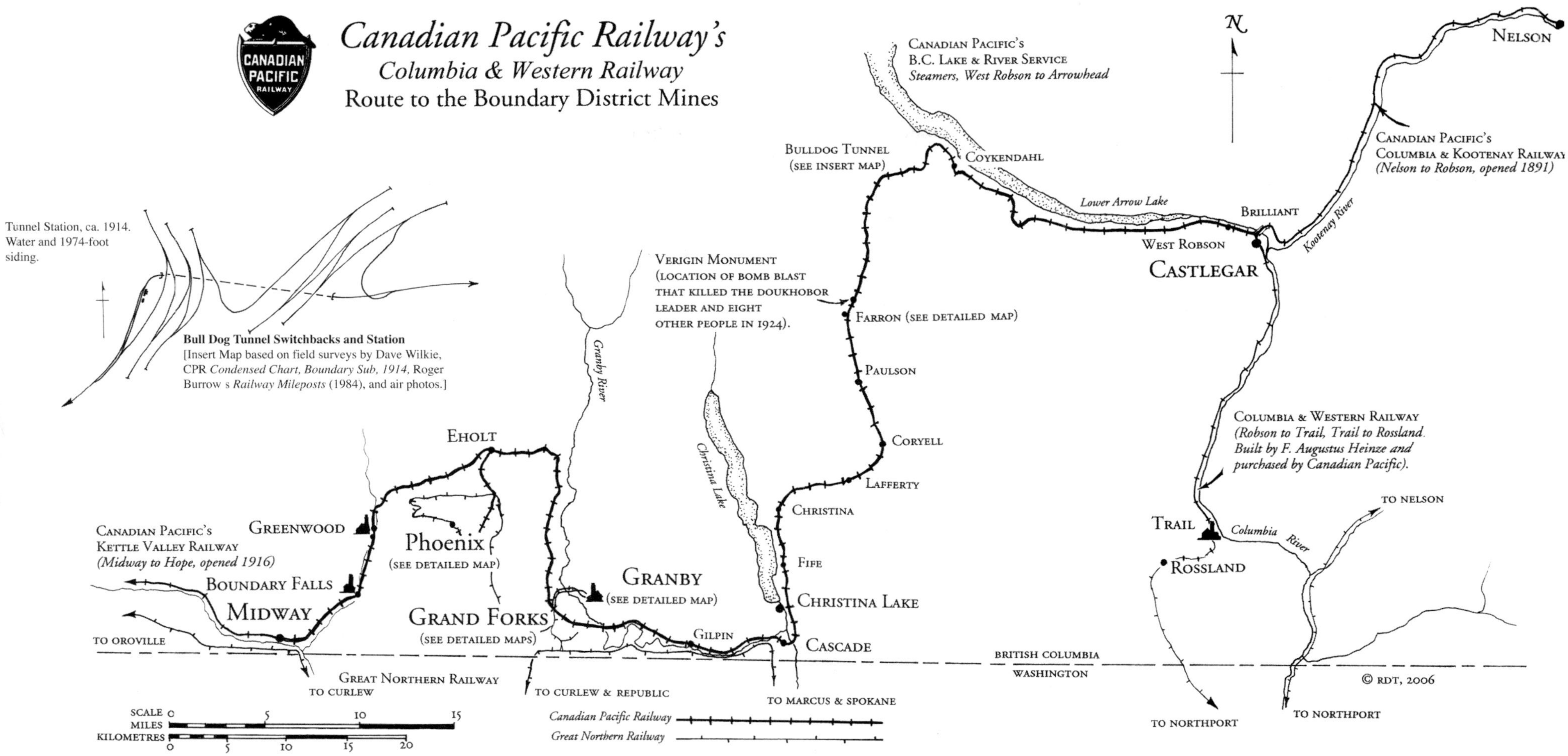

already." He added with a note of triumph and spite, "I am inclined to think that this is a bluff pure and simple, but if they should do so you will notice that we are making them pay pretty well for not giving us a depot here and we offered to build it for them, so it is only showing and demonstrating the vindictive [*sic*] that they have demonstrated towards Grand Forks. …As far as the town is concerned… we would still continue to grow exactly the way we have…."[25] For all the confidence, the CPR was not bluffing and detoured around the city to establish its depot to the northwest of town.

Behind this dispute was rivalry between two townsites each vying for supremacy and land sales. The original location, Grand Forks promoters feared, would give advantage to the townsite to the west. Sullivan suggested running the alternate survey as a bluff and both he and Tye were surprised to get a wire from Montreal instructing them to use this new survey and build around Grand Forks. "I never did find out what influence was brought to bear," reflected Sullivan years later.[26]

In the overall plans, a minor relocation requiring two bridges was not a serious concern for the CPR. Nonetheless, the bridges were substantial and the Grand Forks station location was, for the duration of its various uses as a passenger and freight depot over the next 60 years, an inconvenience. The first bridge required a 150-foot

Tunnels on the CPR's Columbia & Western Ry.

Mile 5.5	194 feet
Mile 12	187 feet
Mile 15	291 feet
Mile 22	3004 feet
Mile 36	329 feet

–*The Railway and Shipping World*, August 1900.

Grand Forks, shown about 1900, was the largest city on the new railway, but land owners and speculators squabbled with the CPR. The railway avoided the business district and built around the city. The Yale Hotel, painted white, is in the centre of the photo just across the North Fork of the Kettle, and the Columbia & Western's second crossing of the Kettle River at Grand Forks is just visible in the photo, far in the distance, directly above the hotel and to the right of cottonwoods.
–ROBERT TURNER COLLECTION.

Railway Inspector Henry B. Smith Was Impressed...

"The location of the railway, from an economical point of view, has been unusually skillfully made and leaves little room for improvement without heavy expenditure. The general course is as direct as the rugged nature of the country would permit."[27]

–HENRY B. SMITH, SEPTEMBER 1899.

The Most Expensive Yet...

"The road is by all means the most expensive the C.P.R. has ever constructed. The country throughout was exceedingly difficult and the engineering difficulties very great. The entire line cost the company in the neighborhood of $40,000 a mile or a total expenditure of nearly $4,000,000 for 100 miles (160 km) of line."[26]

–THOMAS SHAUGHNESSY, *The Railway and Shipping World,* OCTOBER 1899.

(nearly 46-m) Howe truss span and 210 feet (65 m) of trestlework while the second needed two Howe spans totalling 250 feet (76 m) and 1,220 feet (nearly 375 m) of timber trestle approaches.[28]

A government inspection followed the completion of trackwork as far as Grand Forks. Henry B. Smith, an engineer reporting to the provincial Department of Public Works, was requested to inspect the line, and on September 20 and 21, 1899, he travelled with Sullivan and A. C. Dennis, the engineer in charge of track-laying and surfacing. They toured by train from West Robson to the summit, and from there to Grand Forks they rode on a handcar, going downgrade, to give the western slope a more thorough inspection. The next day they reversed the procedure and took the train east to the summit and rode downgrade on railway bicycles, in what perhaps whimsically could be regarded as a preview of the popular rails-to-trails developments a century later.[29]

Smith was impressed with the work and the difficulties that had been overcome, as he observed in his report which is quoted in the adjacent sidebar. He did note, however, that nearly 50 percent of the track lay on curves and the longest tangent was only 5,814.6 feet (1772.3 m) in length. The 29 rock retaining walls covered a combined length of 2,640 feet (805 m) and varied in height from 10 to 40 feet (3 to 12 m) and from 30 to 270 feet (9 to 82 m) in length. Forty-nine timber bridges, which spanned a total of 2.5 miles (over 4 km), were another significant feature of the construction and the high costs. Smith's criticisms were comparatively minor. The Porcupine Creek trestle had been built with 18-foot posts instead of the usual 30-foot size making the 705-foot (215-m) structure 10 decks high, but he did not suggest its replacement. He strongly recommended immediate installation of safety spurs on any grades over 2.2 percent. However, these were never built, probably because the excessive grades were confined to the temporary Bulldog switchbacks over the summit, and they were soon replaced by a tunnel.

Passenger service began on Monday, September 18, 1899, with a scheduled time for the 67-mile (107-km) run from West Robson to Grand Forks over the Bulldog switchbacks of 5 hours and 29 minutes, for an average, rather ambling speed of 13.5 miles (21.7 km) per hour. The temporary switchback route itself took one hour to traverse, and CPR instructions admonished crews that "switches on switchbacks must be kept set for High Line at all times." At first, service was westbound on Monday, Wednesday and Friday and eastbound on Tuesday, Thursday and Saturday. Trains were officially

The station at Cascade, just west of the big trestle over the Kettle River, was typical of the small structures built at communities along the railway. The station agent and his family lived on the upper floor while the ground floor was devoted to the operator's office, a waiting room and space for freight, mail and baggage. The telegraph provided communications along the railway, and operating instructions for the train crews, including clearances to proceed to the next station, and information about where other trains were to pass, were relayed over the wire. This eastbound, with engine No. 403, is one of the first trains through Cascade, photographed on September 19, 1899.
–BC Archives, 45358.

classified as "Second Class Mixed," carrying passengers, mail, express and freight.[30]

Special trains were operated for officials and dignitaries within a few weeks of the opening of the railway to Grand Forks and shortly after to Greenwood. One such train carried Collingwood Schreiber, Assistant Minister of Railways for Canada; Chief Engineer William Tye; Assistant Freight Agent Peters; W. H. Aldridge, manager of the Trail smelter; contractors Timothy Foley, Peter Larson and J. W. Stewart; Dick Porter; and the CPR auditor from Trail, J. W. Kimball. The special ran in the impressive time of just 3 hours and 10 minutes, requiring only 28 minutes to climb and descend the switchbacks that bypassed the incomplete Bulldog tunnel. It then travelled through to Greenwood where it followed the track-laying train into town.[31] Soon after, on another special train, Thomas Shaughnessy toured the line in company of William Whyte, manager of Western Lines; Robert Marpole, general superintendent of the Pacific Division; L. A. Hamilton of Winnipeg, who was responsible for "all townsite difficulties and CPR lands"; J. P. Gutelius of Trail, the superintendent of the Robson–Rossland branch; Captain James W. Troup, manager of the CPR's lake and river steamer services; P. A. Peterson, CPR chief engineer of Montreal; H. C. Cambie, chief engineer of the western division; George McL. Brown, executive agent for British Columbia; Chief Engineer William Tye, returning with his engineering colleague John Sullivan; A. C. Dennis; and W. H. Aldridge, manager of the Trail smelter.[32]

This was Shaughnessy's first trip to B.C. as president of the CPR and his first trip to the Boundary District. Shaughnessy told *The Boundary Creek Times* he was "well satisfied" by his trip. Shaughnessy would not build west of Midway, he explained, because "we have already expended nearly $5,000,000 in this district and we are anxious to secure some return before building farther." Moreover, branch lines and other improvements were still needed.[33] Well satisfied he may have been, but on the return trip the following day his private car derailed on the Bulldog switchback and he declined to continue the journey in it, preferring to ride in the caboose.

Shaughnessy Was Cautious...

"When I thought of the large amount of money we were spending in this Boundary Creek country I had some doubt regarding the wisdom of venturing so much in a new district, but after becoming more intimately acquainted with your wonderful resources... the investment by the CPR will eventually turn out to be a profitable one."
–*The Boundary Creek Times*, October 28, 1899.

Soon after the railway was opened, a passenger train winds its way through the mountains to Grand Forks. At right, another passenger train passes over the long trestle and truss bridge over Boundary Creek just east of Greenwood. Later this structure was replaced by a large gravel fill. A concrete-lined tunnel carried the road through the base of the fill, and it survived the removal of the tracks and earthwork.

–Robert Turner Collection; Greenwood Museum.

3536-44

COLUMBIA & WESTERN RAILWAY CO.,

CONSTRUCTION DEPARTMENT.

TIME BILL NO. 1.

—AND—

SPECIAL RULES,

FOR THE EXCLUSIVE USE AND GUIDANCE OF EMPLOYEES.

TO TAKE EFFECT

Monday, September 18, 1899, at 24:01 O'clock.

(PACIFIC OR 120th MERIDIAN TIME).

W. F. TYE,
Chief Engineer of Construction.
Trail, B. C.

J. S. LAWRENCE,
Trainmaster,
Material Yard.

TRAINS WEST. READ DOWN.		TIME BILL NO. 1	Telegraph Calls.	TRAINS EAST. READ UP.	
Second-class.					Second-class
No. 1. Mon., Wed., Fri. Mixed.	Miles from West Robson.	STATIONS.		Miles from Grand Forks.	No. 2. Tue., Thur., Sa Mixed
9.15	0.0	DE. West Robson AR.	V. G	71.6	13.00
		2.3			
9.25	2.3	Material Yard DE. § AR.	M. A.	69.3	12.47 / 12.17
		9.1			
10.02	11.4	Shield's §		60.2	11.41
		10.2			
10.46	21.6	Tunnel, (East End)	H. I.	50.0	11.00
		2.9			
11.16	24.5	Tunnel Summit		47.1	10.30
		2.8			
11.46	27.3	Tunnel, (West End) §		44.3	10.00
		7.5			
12.16	34.8	Summit §	S. M.	36.8	9.30
		8.4			
12.50	43.2	Gladstone	G. S.	28.4	8.56
		10.6			
13.33	53.8	Sutherland Creek		17.8	8.12
		4.8			
13.52	58.6	Cascade	C. A.	13.0	7.52
		4.8			
14.11	63.4	Gilpins		8.2	7.33
		8.2			
14.44	71.6	AR. Grand Forks DE. §	G. F.	0.0	7.00

§ Water Tanks.

–Les Kosma Collection.

A shortage of rail, and undoubtedly a desire to secure the mining traffic as quickly as possible, led to the spurs to the mines being laid immediately and track-laying west of Greenwood to Midway being deferred until the following spring. Meanwhile work progressed on Bulldog tunnel. By mid-November the headings had progressed over 1,200 feet (365 m) at the east end and passed the 1,280-foot (390-m) mark from the west, leaving just over 500 feet (150 m) to drill and blast. The unsatisfactory nature of the switchbacks was demonstrated again late in November when the westbound passenger train derailed, causing a three-hour delay.

Construction reached Greenwood that fall, and Henry Smith carried out an inspection of that section with John Sullivan on November 21 and 22, 1899. They travelled by train to Eholt Summit and from there took a velocipede down to Greenwood, and the next day they reversed the procedure to complete the tour back to Grand Forks. Smith generally was impressed with the quality of the construction and in particular with the section high above the North Fork Valley where the line pierced two rock bluffs with tunnels 402 and 511 feet (122.5 and 155.8 m) in length. However, he found the curvature, which totalled 53.5 percent of the line, to be poor practice and recommended eliminating some curves. On the 14 miles (21 km) west of Grand Forks the grade stiffened to 2.3 percent and continued for most of the way to Eholt climbing 1380 feet (over 420 m). On the western side the tracks dropped 630 feet (192 m), on a grade of 1.38 percent, to reach Greenwood. Once again, Smith recommended safety spurs on the steep grades but none was ever installed. General practice at the time did not suggest safety spurs on grades until they were in the 4 percent range. His major complaint regarded the falsework on a timber bridge over Boundary Creek: "the falsework erected is of the usual flimsy contractors' pattern and appears totally unfit for any traffic other than that required by the contractor."[34]

Passenger service beyond Grand Forks to Greenwood began on Monday, November 20, 1899, on a schedule that took 1 hour and 30 minutes to cover the 22.7 miles (36.5 km) between the towns. Equipment assigned to the lines west of the Columbia, including the trackage to Trail, consisted of seven locomotives, three passenger coaches, six cabooses, 88 airbrake-equipped flatcars, 40 handbrake-equipped flatcars and 350 boxcars. Initially, all major servicing was done at Trail where offices also were located.[35]

To open the new line more formally, and to promote the new railway and the region with the business community, a gala, three-day excursion was organized for December 7, 1899, the first anniversary of a similar event marking the inauguration of service on the Crowsnest Pass line by the CPR. One hundred and fifty members of the various boards of trade and city councils, newspapermen and representative merchants of cities from the

Kootenays and from Vancouver, Victoria, New Westminster and Nanaimo were invited to attend. The trip through the mountains in the special four-car train was spectacular and the passengers, apparently as many as 170, were treated to receptions and to tours of Cascade, Columbia–Grand Forks and finally Greenwood, which was considered ideally situated to "command the trade of the whole [of the district] when it is developed."[36]

By December 28, 1899, the Bulldog tunnel was holed through with mucking out and track-laying remaining. Although about 500 feet (150 m) of the tunnel had been cut on a 10-degree curve, noted *The Railway and Shipping World*, "the instrument work and chaining was so accurate that the transit lines met within 1/16 of an inch. The levels came out with no appreciable difference."[37] Engineers Tye and Sullivan walked through the bore the next day and on February 12, 1900, the first train passed through the completed tunnel. The saving in time for passenger trains was one hour and five minutes with the tracks passing 434 feet (132 m) below the old summit of the switchback route. For freights, the savings were even more dramatic because they usually had to be "doubled" or taken up over the switchbacks in two sections due to the combination of the 4 percent grades and the short 500-foot (152-m) tail-tracks at each switch. Most of the climb to Farron, the summit of McRae Pass, was now a steady, though compensated, 2.2 percent grade although it lessened to 1.1 percent through the tunnel.

Cascade, a construction camp and boom town on the Columbia & Western, prospered briefly and benefited from the hydroelectric plant, at right in the photo, on the Kettle River's falls that gave the community its name. The power was used at Grand Forks and Greenwood by the smelters, by mines in the area and by the growing towns. The railway is in the middle distance in the photo.

–RESTON PHOTO, WILSON COLLECTION, OKANAGAN ARCHIVE TRUST.

The Columbia & Western crossed the Kettle River on a huge timber bridge composed of long trestle approaches and two substantial Howe truss spans supported by a trestle pier. The massive bridge was 1,500 feet (457 m) long, 90 feet (27.5 m) high, and required 1,500,000 board feet of timber to construct. In 1908 the approaches were filled with more than 400,000 cubic yards (over 300 000 m^3) of gravel. Later six steel girder spans replaced the Howe truss centre spans. This is an eastbound train, photographed on September 21, 1899, headed by Engine 403, a 2-8-0 built in 1886 for the CPR main line in the Rockies. –BC ARCHIVES, 45367.

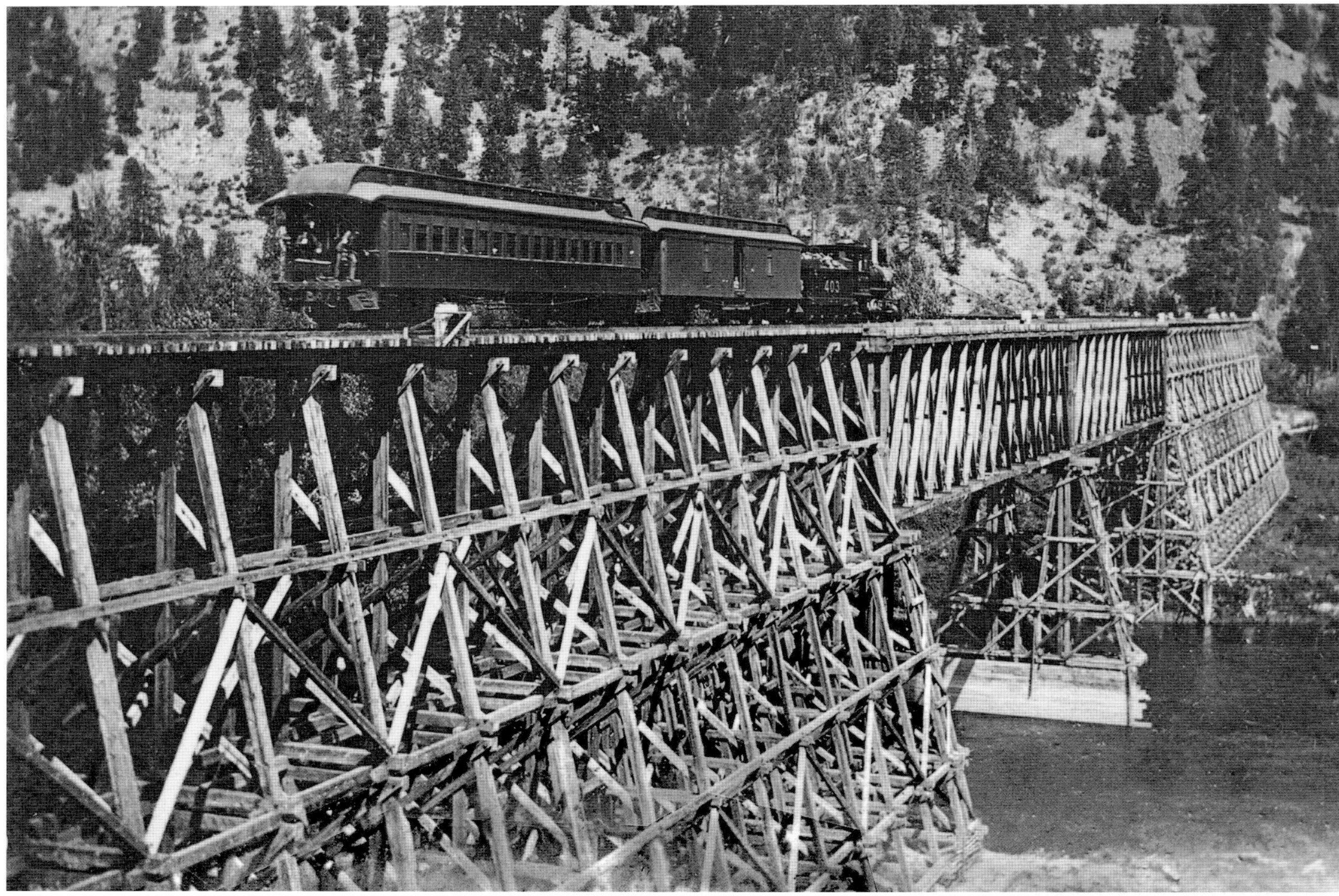

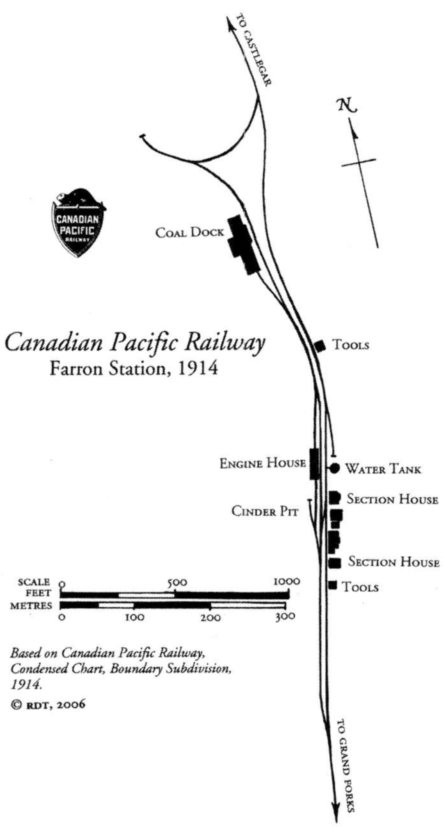

Meanwhile construction was underway on a branch line to Phoenix camp from Eholt, the summit of the divide between Grand Forks and Greenwood. The grade from Eholt to Phoenix was cleared and built almost due south up into the mountains and then turned west and north to reach the mines. Phoenix camp was located in a bowl, much like a gold pan tilted downward towards the northwest. The CPR entered from the east at almost the highest point on the rim which meant that while extensions to the Knob Hill and Old Ironsides properties could be reached by short spurs running slightly uphill, all other extensions to mines such as the Brooklyn and Stemwinder had to run downhill. Consequently, all loads had to be brought uphill over this rim before being eased down to Eholt. At Eholt an important railway junction developed with a station, servicing facilities for locomotives including a water tank and seven-stall roundhouse, a rail yard, bunkhouses and a modest workers' community with the usual stores and hotels.

The gradients on the Phoenix branch would forever be an operating problem for the CPR crews. The line climbed 1,506 feet (459 m) in just 9.7 miles (15.7 km) from Eholt to Phoenix, with grades averaging 3 percent overall, but between miles 5.5 and 7.5 they reached a daunting 4.63 percent. Not only was safety a concern, and two safety switches were installed, on the excessive gradient, but also the hauling capacity and braking ability of the locomotives was severely limited and operating costs were excessive.

On May 21, 1900, the first regular train, with engine No. 312, arrived at the Old Ironsides at Phoenix. Fred Clutterbuck was engineer and W. J. Elliot was fireman.
–BC ARCHIVES, 44272.

In January snow in the higher elevations brought track-laying to a halt on the Phoenix line beyond Hartford Junction at Mile 7.5, and the track gangs were transferred to work on the railway construction underway between Greenwood and Midway, much lower in elevation. The small community just north of the B.C.–Washington border was reached on January 20, 1900. The distance west of Robson to Midway, not including the Bulldog switchbacks, was 99 miles (159 km).

Meanwhile, mining interests were responding to the improvements in transportation offered by the CPR. By early March 1900 the smelter at Grand Forks was nearing completion and the first shipment of coke had arrived from the Crowsnest Pass mines. Grand Forks' *Daily Gazette* reported that 1,000 tons of ore from the City of Paris mine had been accumulated into the first roast heap and that 800,000 feet of lumber had been used in the construction of the dam for generating hydro power across the North Fork of the Kettle River. By this time machinery was being installed in the reduction works and the powerhouse for the smelter.[38]

By mid-April, a spur line had been completed from Greenwood to the Mother Lode Mine near the Deadwood Camp to the northwest. With warmer weather once again in the mountains, the crews returned to complete the trackage to Phoenix. The last few miles were completed quickly and the last spike was driven home at 11:45 a.m. on May 14, 1900. The next week the first regular train entered Phoenix hauled by engine 312 crewed by Fred Clutterbuck, engineer, and W. J. Elliot, fireman.[39] Tracks reached the Old Ironsides on May 23, 1900, amid further celebration.

The CPR soon developed a web of trackage to facilitate ore shipments from the major mines. Spurs were completed to the southeast of Phoenix to the Winnipeg and Golden Crown mines from Hartford Junction, and with trackage laid to the Old Ironsides and Knob Hill mines in Phoenix proper, significant ore shipments could begin at last. On Sunday, May 27, 1900, the first train of ore consisting of four carloads from three different mines, the Winnipeg, Golden Crown and Athelstan, was on its way to Trail for smelting. Then on July 11,

Construction Pay Rates...

"The labourers engaged on track-laying and surfacing of the Mother Lode and Granby Smelter spurs were paid from $2.00 to $2.25 per day. The General Foreman was paid $5.00 a day, and the ordinary Foremen $3.00 and $3.50. These men are working side by side with others whose wages are 10% less. Upon what basis can these high wages be paid?"
–SHAUGHNESSY TO TYE, MAY 31, 1900.

Buffet Car Service...

"The promised buffet car on the Columbia & Western railroad has arrived in the division and will be put into service tomorrow. The car is of the Fournier reclining chair pattern, and is complete in every respect. Meals will be served en route a la carte at reasonable tariffs, and the innovation will be welcomed by passengers bound for the Boundary. The Rossland train will pick up the buffet car daily at Smelter Junction [Trail], and the return train on the same day will connect with it at Grand Forks for the return trip."
–*The Rossland Miner*, JANUARY 7, 1902.

The Columbia & Western's terminal at Midway was a modest affair in the early 1900s with a wooden depot, freight shed and roundhouse surrounded by golden hills. This view is from *The Mining Record* in 1901. Freight wagons still took goods west. At right is the roundhouse just a few years later. At far right is the Granby Smelter in 1900 as it began processing the Boundary ores.
–Robert Turner Collection above and far right; Canadian Pacific Archives, A2650, centre.

The Boundary Falls smelter was taking shape as a major industrial complex by 1901, and, as demands for copper grew and production rose at the mines large-scale expansion followed. The scene shows a short two-car C&W passenger train steaming by the smelter, on a cold winter day.
–Robert Turner Collection.

1900, the initial ore train of five cars, each of 30 tons capacity, ran to the new Granby Smelter at Grand Forks and was met by dignitaries and the music of a brass band. The next day, a second train followed with 150 tons of Knob Hill ore. Because smelting had to await the arrival of more coke from the Crowsnest, the ores were stockpiled temporarily.[40]

Finally on Tuesday, August 21, 1900, the Granby Smelter was ready to process ore. "At 5 a.m., Tuesday," reported *The Daily Gazette*, "the fires were started, [and] between 9 and 10 a.m. the first ore was put in the furnace. At one p.m. the slag was started and by 3.30 p.m. the first matte ever turned out by a Boundary smelter had been produced." By the end of the month, the first shipment of matte, one carload of 20 tons in total, from the Old Ironsides, which produced "99 percent of the ore," with the remainder from the City of Paris, Lone Star and Washington mines, was sent via the CPR to the east for further smelting and refining. It was assayed at about half pure copper, with traces of silver and gold. Initially only two furnaces were in operation at the Granby Smelter, but equipment was continually improved, reducing the impurities in the copper matte, and more furnaces were added to keep up with the production of ore from the Phoenix mines and others at Republic.[41]

CPR freight rates for hauling the Phoenix ores to Trail had been $7.00 per ton, but with the shorter distance to the new smelter, the freight and treatment rate for the Phoenix ores at Grand Forks was only $3.50 to $4.50 per ton, putting the production on a much more competitive basis. Moreover, with this dramatic reduction in costs, lower grade ores could be mined and smelted profitably. Although the CPR's rates fell, it more than made up its revenues by the increase in volume of ore, in the shipment of coke and coal from the Crowsnest and the rise in general passenger and freight traffic.

The Mines Branch of the Federal Department of Mines reported enthusiastically on the progress in the district and the completion of the railway. "Among the railway lines constructed—whose operation has contributed largely to the prosperity of the mining industry during recent years—may be mentioned, the British Columbia Southern, which opened up the valuable coal fields of the Crowsnest; and the Columbia and Western road, between Robson and Midway, whereby access to the Boundary District was attained. By the completion of the former, an abundant supply of excellent coke, at less than one-half the previous cost, was at once made available for metallurgical purposes; while the completion of the Columbia and Western line was the means of opening up what is now the largest copper producing area in Canada."[42]

The new railway and the construction of the smelters were keys to the development of a lucrative copper industry in the Boundary District, but other considerations also were critical. As the Mines Branch later noted, "four general conditions have contributed to the successful mining and smelting of the low grade ores; the absence of any one of which would render their working nearly,

Greenwood was not content with being a mining camp, and was incorporated as a city in 1897. By 1903, when this photograph was taken looking somewhat southeast, it was a bustling community. By 1905 the population was recorded at about 1500 and *Henderson's British Columbia Gazetteer and Directory* for that year listed an impressive range of businesses and services in Greenwood and Anaconda (as the smelter camp was called). Included were three banks and three breweries, two hospitals, grocery, stationery and general merchandise stores, a butcher, a coal and wood dealer, an ice dealer, shoemaker, dentist, doctor, baker and at least 19 hotels and rooming houses, often providing lodging for the many single men who worked in the mines and smelter. As well there was the Greenwood Miners' Union, No. 22 of the Western Federation of Miners, the Greenwood Lodge of the Knights of Pythias, the Boundary Valley Lodge, No. 38 of the Independent Order of Odd Fellows, and the Greenwood Lodge, No. 22 of the Masons and, of course, a newspaper, *The Boundary Creek Times*, Mr. Duncan Ross, managing editor. In the foreground of the photo is the CPR line to Grand Forks and the station. Long after the boom, Greenwood boasts of being the smallest city in Canada.

–Wm. Notman & Son, No. 3683, McCord Museum.

quite unprofitable… (1) the enormous extent of the ore bodies, making mining on a huge scale possible; (2) the solidity and firmness of the ground which makes the opening of great stopes, without the use of timbers, not only feasible but safe; (3) the chemical composition of the ores, which renders them self-fluxing, and obviates the use of barren flux; and (4) the proximity of the district to the vast coking-coal fields of the Crowsnest Pass. Other contributing aids are: abundant electric power, plentiful water and timber, dry mines, and an unsurpassed climate."

Once the railway was in operation, there was still much work to be done in improving the line and replacing temporary works. Besides the Bulldog tunnel, the other major work needed was a bridge over the Columbia River at what is now Castlegar, where the crossing of the river between Robson and Robson West was a major obstacle in the Canadian Pacific's route to the Boundary District. In the first few years of construction and early operation, rail cars were ferried across the river by the steamer *Lytton* that moved barges between Robson and Robson West. Although much of the traffic to and from the Boundary was moving via the Arrow Lakes

BRIDGE

Castlegar's unusual station was built in the middle of the wye which divided the tracks between the new railway to the Boundary mines to the right and the branch line to Trail and Rossland which diverged to the left. An appropriation of $5,172 was authorized on March 10, 1902, for the station by CPR President Shaughnessy. In September another $828 was approved for the platform. In the background are special cars used for carrying coke from the Crowsnest mines to the smelters. The station survives as Castlegar's beautiful community museum and tourism information centre.

– J. W. Heckman, Canadian Pacific Archives, A2200.

steamer route, the need was clear for a direct rail link eastward through to Nelson and the connections with the Crowsnest Pass railway. Contracts were let for an impressive bridge over the Columbia at Sproat's Landing about 1.5 miles (2.5 km) south of Robson. John Gunn of Winnipeg won the contract for the foundations and piers, and H. W. D. Armstrong was the resident engineer in charge of the construction. It became an enduring landmark on the Columbia.

Often temporary bridges were built with the intention that they would be replaced in a few years with permanent structures, but this bridge was built to last. The depth of water at the bridge site varied from a minimum of about 25 feet (8 m) at low water to 35 or 40 feet (10–13 m) during the peak of the spring runoff. Beautifully crafted masonry piers and abutments were built to support, from west to east, a 55-foot (16.75-m) girder span, the two 204-foot (62.2-m) steel truss spans, a swing span providing openings of 54 feet (16.5 m) on each side and two more 55-foot deck girders that completed the structure. The swing span was necessary to permit continued steamer navigation on the river although, after the construction of the railway between West Robson and Trail, large vessels seldom operated south of Robson. On the east side an approach of 2000 feet (610 m) was needed including a 1400-foot (425-m) trestle that was filled almost immediately with material excavated on the western approach near the Castlegar station. The bridge was opened in March 1902.[43]

The CPR spent some $4,500,000 to complete this very rugged piece of railway into the Boundary District, mainly to move the ores from the new mines to their recently purchased Trail smelter and at the same time supply the cities, mines and smelters of the mining communities with the large volumes of coal, lumber and other products they needed. With coal mines developing in the Crowsnest Pass, the possibility of supplying the mines and smelters with CPR-hauled coal and coke was particularly promising. Every 10 tons of ore required at least one ton of coke for smelting, and coal was needed for many power plants in the mines and smelters, although the smelters used hydroelectric power whenever available. In this way, loaded coal and coke trains were operated westbound to the mines, and ore, matte or blister copper was shipped east. Passenger trains carried miners, businessmen, families and visitors to the rapidly growing communities in the Boundary District. However, other smelters were under construction and other railroads were developing plans to tap the massive tonnages of ore in the region. The CPR's monopoly in the Boundary was nearing its end and a new era of intense competition was about to unfold.

On the facing page...

Canadian Pacific's almost finished bridge over the Columbia provided a direct rail connection between its Columbia & Kootenay trackage from Nelson and the Columbia & Western. The beautiful bridge, set on masonry columns, had a swing span to permit sternwheeler navigation. The steamer is the *Lytton*, used until the bridge was opened to ferry passengers and railcars on barges across the river. Workers are finishing the stonework while the water level remains at a winter low. The rotating mechanism, hand powered, was worked from the centre platform.

–William Hall, City of Vancouver Archives, CAN P174.

More Railways for Grand Forks and Republic

Tracy William Holland was an ambitious and aggressive promoter of Grand Forks real estate and the Kettle River Valley Railway, and briefly a mayor of Grand Forks. His dreams of profiting in land sales and developments, and in challenging the Canadian Pacific and the Great Northern for the mining traffic, crumbled as the large companies dominated the mining districts.
–Boundary Museum.

Golden Spikes and Hot Air Do Not a Railway Make...

"If you think it advisable, you may announce that the Great Northern Ry. and Mackenzie and Mann are equally interested in the V. V. & E. Ry. Co. stock."

—J. J. Hill and Wm. Mackenzie, New York, March 21, 1901, quoted in *The Railway and Shipping World*, April 1901.

"Grand Forks jeweler A. D. Morrison has had 50 stickpins made up as gold spikes representing the last spike on the Kettle Valley Lines. All were sold out within two hours of the store's opening."

—*The Grand Forks Miner-Gazette*, April 12, 1901, on the day of the Kettle Valley Lines' opening excursion.

"When the R&GF railway was undertaken it was dubbed the "Hot Air" line because its local promoters had so much trouble in getting down to actual construction. The people of both places are now jubilant because the "Hot Air" line has more men at work than the "Jim Hill" line, is paying better wages and promises to be completed this coming winter."

—*The Phoenix Pioneer*, September 6, 1901

Equipment for the Hot Air Line....

"The construction of the roundhouse for the Republic & Grand Forks on the Ruckle addition was commenced on Tuesday... there are 30 ore cars at Grand Forks and an additional 30 ore cars will be hired from the C.P.R.... Passenger coaches are on the way, and also two large engines each weighing 100 tons and built by Cooke..."

—*The Grand Forks Miner-Gazette*, January 25, 1902.

By the turn of the century, Republic, Grand Forks and the Boundary District became the focus of more railway construction as the mining boom took hold. The first of two serious schemes was promoted by a group of businessmen, principally in Spokane, involved with the Grand Forks Townsite Company and associates. The Washington section of the railroad, which was to connect Republic and its mines with Grand Forks, was called the Republic & Kettle River Railway and was incorporated on August 24, 1900, with a capital stock of $1,000,000.[1] Incorporation of the line on the Canadian side of the border was a bit more difficult and was strongly opposed, not surprisingly, by Canadian Pacific interests. This section was initially called the Grand Forks & Kettle River Railway with capital listed at $200,000. The directors on the Canadian side were James R. Stratton, Provincial Secretary in the Ontario government; Thomas Patrick Coffee, manager of the Trusts & Guarantee Company, Toronto; Frederick Montford Holland, manager of the Dominion Permanent Loan Co., Toronto; George Henry Cowan, barrister from Vancouver; and Tracy William Holland, manager of the Grand Forks Townsite Company. These were some of the same individuals who had tried to bargain with the CPR over the location of the station site in Grand Forks. Argument and negotiation over the location of stations and routes through the city and surrounding lands were debated locally, causing considerable acrimony between the railways and local interests and not surprisingly between the competing railways themselves.

Facing page...

The Spokane & British Columbia's No. 2 switches ore cars up to their main line from the mining spurs in Eureka Gulch near Republic. The photo was taken from the Great Northern's track on their spur line to the Tom Thumb mine. The view is looking south down the gulch at the Lone Pine and Surprise ore bins. The silver-gold ore will be taken to Grand Forks for smelting.
–Dave Wilkie Collection.

Shaughnessy Was Upset by the GNR Invasion...

10th July, 1901.
Dear Mr. Hill,
We now have definite information that a Company controlled or assisted by the Great Northern are about to construct a railway from Marcus to the International Boundary Line, there to connect with the Vancouver, Victoria & Eastern Railway, whose Canadian charter, you have stated, is owned or controlled in whole, or in part, by the Great Northern.

As the construction of this line on the Canadian side involves the invasion of territory tributary to our line between Cascade and Grand Forks, it will be in direct contravention of the understanding that has existed for two or three years between your Company and ours, and I presume may be accepted as a declaration by Great Northern that that understanding is at an end.

Yours very truly,
T. G. Shaughnessy,
President.

The provincial bill authorizing the charter for the proposed railway squeezed through in the final hours of the ninth legislative sitting in Victoria on September 1, 1900. The bill came before the legislature at 4:35 a.m. and was supported by a majority of the members. In speaking for the bill one of its supporters pointed out that it was backed by capitalists, not charter-mongers or speculators, and would bring Republic ore to British Columbia. "Would [the legislature]," he asked, "follow the CPR or the people? Which has the most influence in British Columbia?" It passed 17 to 15 and the legislature closed at 6:15 that morning.[2]

Whether capitalists or speculators were backing the project might have been a debatable point, because there certainly were speculators in the Grand Forks townsite schemes. The company backers did not rush construction and it became known locally as the "Hot Air Line." In fairness, there were many problems in beginning work. The Republic & Kettle franchise held by Warren Miller and associates was finally purchased in mid-1901 by an American lawyer from Republic, W. C. Morris. Born in Iowa in 1870 and admitted to the bar in 1891, he moved from Seattle to Republic in 1898 and became associated with Tracy Holland two years later. In the end, the efforts of these two men pushed the scheme ahead. On the Canadian side, the railway received a federal charter on May 23, 1901, as the Kettle River Valley Railway and, when amalgamated with the Republic & Kettle River Railway, the combined operation became known as the Kettle Valley Lines.

Because the southern section of the line had to pass through the Colville Indian Reservation, which was administered by the Department of the Interior, Holland and Morris had to make several trips to Washington, D.C. Finally, the order authorizing the building of the railway was signed by E. A. Hitchcock, Secretary of the Interior. Reflective of attitudes of the times, it appears no one cared how the Colville Tribes felt about the railroad being built through their reservation, any more than their wishes had been considered when the reservation was opened to mining and, soon after, white settlement. North of the border, the situation was administratively different, but the result much the same; few First Nations' concerns were considered in land questions relating to railway and mining developments.

While Holland, Morris and their associates were preparing to build their railway, a much greater force, in the form of James J. Hill's Great Northern Railway, was also setting its sights on the riches of the Boundary District and Republic. Moreover, Hill had plans for a much more extensive expansion to the west including more branch lines into the lucrative mining regions in British Columbia. Great Northern trackage had already reached the Pacific coast at Everett, and GN established its major terminals at Seattle on Puget Sound. Hill had acquired connections, as noted earlier, through subsidiary companies or at least ones in which he and associates or the GN had a controlling interest, north to Brownsville across the Fraser River from New Westminster, and had his sights set on a direct connection into Vancouver. As an interim measure, pending completion of a railway bridge across the Fraser, a ferry was used and passengers could ride the New Westminster-to-Vancouver electric interurban service initially provided by the Westminster & Vancouver Tramway Company, which eventually became part of the British Columbia Electric Railway system.

Hill used the charter of the Vancouver, Victoria & Eastern Railway & Navigation Company, the VV&E, on the Canadian side of the border for his system of branch lines and a through route across the southern districts of British Columbia. With connections in Washington State, this allowed Hill to develop a Marcus-to-Republic route that had been denied Corbin, as well as expand to the west. Hill's direct access to Vancouver came after 1904 with the completion of the railway bridge over the Fraser River at New Westminster and the construction of the Vancouver, Westminster & Yukon Railway, from New Westminster to False Creek in Vancouver.[3] This railway was financed and then taken over by the GN, before being transferred to the VV&E in 1908.

Great Northern engineers surveyed a water-level

route from Marcus to Republic, coming up the Kettle River to cross the international border at Laurier, near Cascade, then west for 15 miles (24 km) on the Canadian side of the border still following the Kettle upstream, then back to Curlew, 10 miles (16 km) south of the border, along Curlew Lake and finally through the canyon of the San Poil River to Republic. Almost all of the mines at Republic were up Eureka Gulch west of town and a 9.8-mile (15.8-km) extension was required. These mines included the Quilp, Lone Star, Surprise, Mountain Lion, Tom Thumb, Black Tail, Ben Hur, San Poil, Trade Dollar and Morning Glory.

The Great Northern surveyors had managed to get the jump on the Kettle Valley Lines' crews and had filed surveys of a line with grades no greater than 0.6 percent from Marcus to the south end of Curlew Lake. For the remainder, a maximum of 1.5 percent was required on the last few miles above the San Poil canyon. The line hugged the west shore of Curlew Lake for its entire length, forcing the Kettle Valley Lines' surveyors to twist their line up along the rolling benchland west of the lake for a backdoor entry into Republic over 3,000-foot (900-m) Barrett Creek Pass and the west fork of Granite Creek. This entailed a nine-mile (15-km) climb from the north end of Curlew Lake with a two-mile (over 3-km) detour up Trout Creek to gain elevation while keeping the long grade to 1.5 percent. At the summit of the pass, the rails were 700 feet (nearly 215 m) above Curlew Lake. It was fortunate that the northbound grade from Republic to the pass was only 0.8 percent because the loads of ore would be hauled in that direction.

Contracts for the 73.4-mile (118.1-km) GN line were let in July 1901 at a time when the survey crews were still mapping out the final route. At the same time, the GN's right-of-way agent, E. Charles, was working with Colonel A. M. Anderson, Indian Agent for the Colville Indian Reservation, trying to settle claims for the various Indian allotments through which the line had to pass. This proved to be a lengthy process because, as noted, the route had to be filed and cleared through the Secretary of the Interior in Washington, D.C.

The chief contractor for GN's Marcus–Republic line was Patrick Welch, the experienced railway builder who had worked on the CPR's line into Grand Forks. There were many other familiar names in railway construction among the subcontractors including Burns & Jordan, to whom went sections requiring tunnelling or heavy rock work, and the Porter Bros. who handled the bridge-building.[4] Other subcontractors included the Twohey Bros., well-known western railroad contractors; Johnson & Peterson; and L. Wilson & McClearan & Harris. Culverts were assigned to a contractor named Van Norman.

The route included two major bridges. The first spanned the Columbia River near Marcus and the second crossed the Kettle near a station called Boyds, about three miles (nearly 5 km) above its confluence with the Columbia. The Columbia River crossing was 1,250 feet (381 m) in length, with a massive centre span of 480 feet (146 m). At the time of its construction it was considered to be one of the longest, heaviest and most expensive structures of its kind in the Pacific Northwest. In later years the construction of the Grand Coulee Dam and the backing up of the Columbia to form Lake Roosevelt required a relocation of this trackage and a new crossing of the Columbia several miles south of this original route.

The autumn of 1901 saw two rival railway companies in a race between Grand Forks and Republic with the CPR well established in the district and watching from the sidelines, but clearly concerned about the outcome. If the construction of the Kettle Valley Lines was a worry, the spreading competitive threat of the Great Northern was a serious concern to the Canadian Pacific and a reflection of a growing rivalry between the two major railway companies.

With the Great Northern already ahead on the ground and with its contracts let, the Kettle Valley Lines finally completed its legal hurdles and awarded contracts late in August 1901 to Charles P. Ferguson of Spokane for the construction of the entire line except for a few miles already graded. The 35 miles (56 km) of railway

James J. Hill Replied...

August 3, 1901

T. G. Shaughnessy

Dear Sir:

In regard to the extension of our line up Kettle River, allow me to say that we do not wish to invade Canadian territory in any manner beyond what is physically necessary in order to get access to the country south of the International Boundary. We have given time and money trying to get a line from our Spokane Falls and Northern to the mines at Republic and elsewhere south of the Boundary, but the mountain ranges make the physical conditions such that the only line with a low grade is by following the River. This carries us across the Boundary for a distance of fifteen miles, where we follow the River back into the United States. We have no wish to invade the country served by your lines.

Yours truly,

J. J. Hill

President.

The Yale Hotel was one of Grand Forks' finest, and often hosted banquets and special events. The engraving shows the hotel, looking southwest across the North Fork of the Kettle. At the time it was said to be the largest hotel in the interior of B.C., and was owned by John Manly who headed a group called the Yale Hotel Company. Unfortunately, it was completely destroyed in the Grand Forks fire of July 11, 1908. A new Yale Hotel, somewhat smaller, was built to replace it. The Yale Hotel remained a landmark of Grand Forks long after the Kettle River Valley Railway, the Spokane & British Columbia and other railway ventures had been all but forgotten. Known for its fine Doukhobor food, it survived until the early 1990s when it was destroyed by another fire; fortunately, Grand Forks' reputation for excellent traditional foods remains.

–Dave Wilkie Collection, from *The Grand Forks Miner-Gazette,* April 6, 1901.

Undeterred, the GN entered Canada just south of Cascade in early March and, to speed up construction, built temporary track around a large outcropping where a tunnel was planned. Their track-laying machine, with its large crew, was putting down two miles of track a day as they worked along the south bank of the picturesque Kettle River. Work came to a halt only when they reached the disputed Manly property. Blocked by deputies, GN progress halted until April 5, 1902, when, with the help of members of the Board of Trade, the railway purchased the entire Manly property.

Although still incomplete, and caught up in its squabbles with the Great Northern, on February 8, 1902, the Kettle Valley Lines began hauling ore to Grand Forks from its railhead south of Nelson (Danville). Sacked ore, 150 tons of it, was brought north by teamsters from the California Mine south of Torboy, destined for the Granby Smelter at Grand Forks. On March 13, passenger service was started as far south as St. Peter's Creek, where a station called Pelham was established. The train left Grand Forks at 7:30 a.m., returning in time to connect with the eastbound CPR train. Passengers travelling from Republic to Grand Forks had to endure only 12 miles (about 20 km) of bumpy stagecoach ride instead of the 35 wrenching, bone-jarring miles (56 km) before the coming of the railway. Track-laying continued over Barrett Creek Pass at a rate of about three quarters of a mile (about one km) a day. Eight trestles, including four 2,200 feet (670 m) in length, were required on this section.

On April 12, 1902, Kettle Valley Lines' track was nearing Republic and a formal opening, with a special excursion from Grand Forks, was arranged, accompanied by the Grand Forks Cornet Band. And although the grand opening turned into something of a fiasco, it nonetheless ended on an enjoyable note.

"Instead of getting away at 9:00 a.m. as was planned, the excursion was delayed until long after 10:00 to allow promoters of the line and the crowd to be extensively photographed. Finally a start was made and despite the roughness of the roadbed good time was made until a point near Curlew Lake was reached. Here there was a slight delay to switch around a material train and then the party proceeded once more. A few miles up the line just a short distance below Ferguson the special excursion train met with a work train coming down the hill to dinner and it was only by the luckiest kind of accident that a serious collision was averted. As it was, one man who got frightened and jumped off the work train broke his leg and others were severely shaken up and bruised. The 'Special' then backed down and waited till the men on the work train which followed it down the hill had had their dinner; a delay of about one hour. After the track gang had eaten their meal the party proceeded up the line for a few miles beyond Ferguson where a section of freshly laid and very shaky track was encountered and it was necessary to have the work train couple on and haul the cars of the 'special' over this. The track not being in condition to stand the weight of the heavy engine.

"Finally, the railhead was reached at a point some distance this side [north] of Tom Thumb hill and the party was there loaded into lumber wagons for a five or six mile drive into Republic over one of the roughest roads that could be imagined. Many of the group ended up walking the greater part of the distance, not being able to stand the jolting of the wagons. After a sandwich lunch at Republic the crowds were again marched out of town to the hillside opposite "Old Town" where the farce of "driving the last spike" was acted out. The spike appropriately gilded for the occasion being driven at a point where a couple of rails had been laid to mark what will be the end of the line when it is finished. This bit of comedy was partaken in by Tracy W. Holland, General Manager, and W. C. Morris, Attorney General [*sic*]. Then there was a scramble for the wagons and a rush to get to the end of steel before darkness fell as it would have been practically impossible to get over the road through the swamps after dark. When the party arrived at railhead it was just dark and a start was made at once. The trip required from 6:30 until 11 p.m. Upon arriving at Grand Forks once more the crowd was rushed to the Yale Hotel where a banquet was served. This was

Great Northern 4-4-0 No. 290 brought the first Great Northern Railway train into Grand Forks. The GN border crossing station at Danville, with its small depot, section buildings and classic water tank is at left.
–BC Archives, 44187; and Boundary Museum, 002.488.001.

the one redeeming feature of the entire affair."[8]

The Kettle Valley Lines' backers were clearly trying to keep up the momentum of their brief lead over the Great Northern and to retain the support of the local community. However, it was apparent that the citizens were divided in their loyalties, most people preferring healthy competition over a monopoly, and a delegation to Victoria succeeded in placing the GN's problems before the provincial government. The Attorney General immediately withdrew the injunction blocking progress on the trackwork. The May 3 issue of Grand Forks' *Miner-Gazette* reported "the VV&E [Great Northern] is rushing ahead as fast as possible and it is expected they will reach Republic not later than July 1st."[9]

Troubles continued to plague the Kettle Valley Lines. In a story with the unflattering introduction "Toy railway is slumping and its bridges washed out," the *Miner-Gazette* gave the dreary summary presented at right.[10]

During that winter and hectic spring of 1901-1902 James J. Hill had been busy buying shares of the Granby Smelter on the open market. By early May he had acquired a substantial minority interest in the expanding smelter. So sure were the Great Northern officials that the various injunctions would eventually be lifted—and in this they were correct—that they went ahead with grading north of the Kettle Valley Lines' tracks. They also contracted with the Porter Bros. to build a bridge across the Kettle River near Columbia, a small community west of Grand Forks proper.

The Great Northern trackage in B.C. was inspected and approved for service in mid-June by H. E. Forrest, the federal government inspector, who declared it to be first-class construction. As always, James J. Hill built his railroad branches to main line standards without subsidies. The first carload of freight billed to Grand Forks over the GN was a load of blasting fuse delivered to W. K. C. Manly at the wye on the Cooper ranch south of town. By the middle of July, Great Northern contractors Stewart & Welch had completed a spur to the station at Columbia and on July 29, the GN finished track-laying all the way from Marcus to Republic. The only heavy construction on the line south of Grand Forks had been a short 113-foot (34.5-m) tunnel through a rock spur about two miles north of Curlew and a 770-foot (235-m) long, low trestle of 54 bents across the upper end of Curlew Lake.

Regular passenger train service began over the scenic Great Northern route between Spokane, Grand Forks and Republic immediately after track-laying was finished on July 29, 1902. Although the trip still took the best part of a day, it was a dramatic improvement for the residents of the remote mining communities. Trains left Spokane at 9:45 a.m., and arrived at Grand Forks (actually the Cooper wye south of town) at 3:45 p.m. and Republic at 5:30 p.m. This was a daily service with the eastbound train leaving Republic at 10:00 a.m., arriving at Cooper

The Hot Air Line Had Troubles...

"The road that was going to haul the mines of Republic right down to our doors and smelt the ore by the 'hot air' process is having all kinds of trouble.... Its flimsy bridges being washed out by the slight rise of the Kettle River, its grades slumping and letting the track into the water and its operations being stopped.... The fun started on Tuesday [May 6, 1902] when the bridge over the Kettle River at Nelson, Washington, gave way and four spans aggregating 80 feet [24.5 m] in length washed out. By a lucky chance the passenger train from Republic had stopped to await the clearing away of logs that were jamming on the pilings before crossing.... On Thursday evening the Curlew Bridge was so far out of line that it was deemed advisable to blow out several spans in order to possibly save the balance. This leaves the road with only the little No. 1, jocularly known as 'The Tin Whistle.' "
–*The Grand Forks Miner-Gazette,* May 10, 1902.

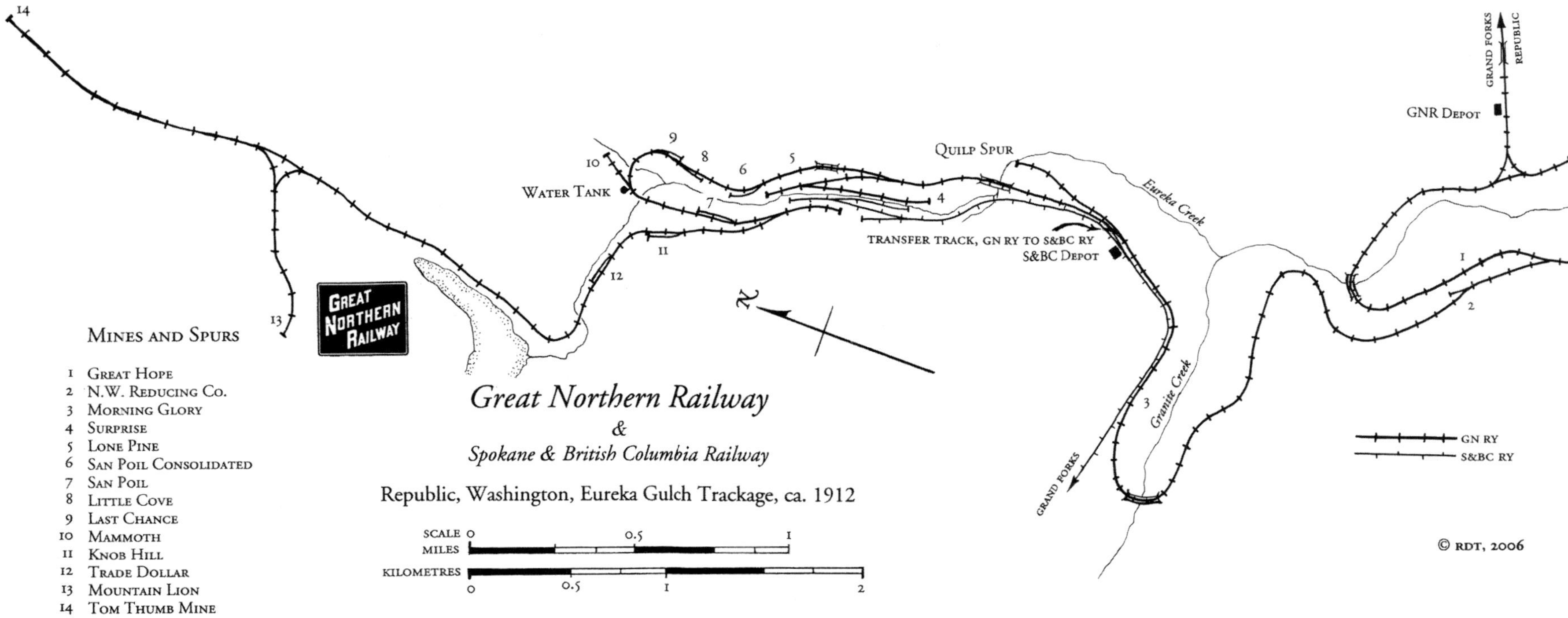

Frontier hotels filled to the rafters when railway construction started. This is Grand Forks' Hotel Colin, opposite the Great Northern station, and, as the hotel advertised, "conveniently located for railway men... [with a] fine line of Wines, Liquors and Cigars always in stock at the bar." –Wilson Collection, Okanagan Archive Trust.

An early Great Northern passenger train at Republic. The four-car train could be the one that took James J. Hill on a tour to Grand Forks and to the camp on a tour of inspection of his expanding interests in 1902. –Dave Wilkie Collection.

Republic, just south of the international boundary, boomed in the late 1800s when parts of the Colville Indian Reservation were opened to settlement and mining. Promising gold and silver deposits brought miners and businessmen to the region, but production quickly fell off all too soon after the railways reached the camp, which struggled on but never reached the heights extolled by the district's early promoters or needed to make the railways prosper and recover their costs. The Spokane & British Columbia's losses grew and it eventually collapsed. The Great Northern had the reserves to cover its losses and shifted its focus to Phoenix and the lucrative traffic ore traffic from its mines. –Frank Palmer, Peltier Collection, Eastern Washington State Historical Association, L84-327.332B.

wye at 11:35 a.m., and reaching Spokane at 6:20 p.m.

Just two weeks before the Great Northern began service to Republic, the Kettle Valley Lines had rebuilt their bridges at Nelson (Danville) and north of Curlew and reopened their line to traffic. On July 12, 1902, the first shipment of ore directly from Republic over the Kettle Valley Lines was delivered to the Granby Smelter via the CPR's spur line. Although just two carloads of ore from the Quilp Mine were moved, it was a beginning. Two more followed on the 14th, and four were delivered from the Surprise–Lone Pine Mine on the 20th. Finishing touches and spur lines to the mines followed, including an 1,800-foot (550-m) spur to the Quilp Mine, a 600-foot (180-m) line to the Lone Pine and Surprise ore bins, a 700-foot (215-m) spur to the Zela M. Mine and a 100-foot (30-m) track to the Blacktail Mine. An 800-foot (244-m) siding was installed at the Republic depot and a 1,600-foot (488-m) track was extended from the CPR crossing in Grand Forks to the Kettle Valley Lines' roundhouse. Despite growing freight shipments, the company did not extend passenger service to Republic until December 8. Trains left Grand Forks at 7 a.m. and reached Republic two hours later. The return trip departed at 1:00 p.m. and reached Grand Forks at 3:00 p.m., in time for passengers to transfer to the CPR passenger train heading east to Castlegar and Nelson.

When the Great Northern Came to Town...

"Hundreds of railway laborers arrived this week and were soon distributed along the route of the proposed Great Northern line between here and Phoenix. The streets have assumed a very lively appearance. The hotels and restaurants uptown and downtown are doing a rushing business.... Phoenix is a lively centre as several hundred men have gone into camp there. There will be heavy rock work in that vicinity."

–*The Grand Forks News-Gazette*, June 4, 1904.

On November 10, 1902, the Kettle River Valley Railway parked its Tenwheeler No. 3 across the Great Northern to block access by its rival. The letters RGFRR were added to the photo, probably by the photographer who sold prints of this picture. They were for the Republic & Grand Forks Railroad, a local name for the line. The two 4-6-0s, with 21x26-inch cylinders and weighing 215,000 pounds (97 522 kg), were said to come from the Cooke works, but details of all three KV Lines engines are unclear, as is their later history; most likely they were scrapped in the early 1920s.
–BC Archives, 44204.

Endless Injunctions...

"Referring to the litigation between the K.V.L. and the Vancouver, Victoria & Eastern Ry. in connection with their respective lines between Grand Forks, B.C., and Republic, Wash., T. W. Holland, General Manager, recently stated that one company had a record of 22 injunctions against its opponent, while itself was the authority for 19 counter injunctions."
–*The Railway & Shipping World*, August 2002, p. 264.

James J. Hill inspected his new railroad as far as Republic and visited the Granby Smelter and Grand Forks where he addressed community business leaders at the Yale Hotel on August 4. The smelter, he commented, was the most modern he had ever visited. He went on to reassure the guests of his plans to build west through the mountains to the coast, but, he emphasized, he did not want "to build a road consisting of a series of toboggan slides. They had to seek and follow the lines of least resistance and low grades." He went on to underscore his disapproval of bonuses and subsidies.[11]

Various injunctions still kept the Great Northern from completing expeditiously their essential branch line into Grand Forks and to the Granby Smelter. Months dragged by and finally on November 1, 1902, *The Grand Forks News-Gazette* announced that the Railway Committee in Ottawa had authorized extension of the VV&E to Granby. "Bonfires blazed here and in Columbia Tuesday night in celebration of the judgment... granting the application... to cross the Kettle Valley Lines and the CPR tracks near here, and build a four-mile [6.5-km] spur to the Granby Smelter as well as an extension of the main line to Phoenix." It went on to note that the contribution of the traffic of coke, smelter matte and other commodities was expected to be $600,000 a year.

One more dramatic, and ultimately foolish, altercation was in store before construction came to an end. This one concerned the GNR crossing the Kettle Valley Lines' tracks en route to their station at Columbia on the outskirts of Grand Forks. Apparently, the Kettle Valley Lines' officials were unaware that an agreement had been reached to allow the crossing in exchange for the cancelling of a similar legal blockade by the GN over the KV Lines' crossing over the GNR at Pelham flats five miles (8 km) south of Curlew.

On Sunday, November 9, 1902, a GN work train came up from Marcus with a gang of men under foreman Duffey with all the material necessary for the crossing including the diamond, ready for installation. Work in cutting the rails of the Hot Air Line's track and in putting in the crossing began at 5:00 p.m. and was finished within an hour. So quickly was the work done that not a person in Grand Forks knew what had transpired until about eight that evening. On Monday morning, Hot Air Line officials learned the news and immediately sent engine No. 3 with a boxcar and flatcar out to block the crossing. The engine was positioned right over the diamond and remained there until 5:00 p.m. as darkness settled over the valley. Throughout the day Great Northern crews had to carry supplies around the crossing and move rail and other equipment to the end of track by handcars. Finally, late in the afternoon, a telegram was received from officials in Vancouver and it was only then that the Kettle Valley Lines' train was withdrawn from the blockade.[12]

At last, the Great Northern was able to build its trackage to the smelter and deliver ore directly from the mines at Republic. Construction of a branch line to Phoenix still remained to be completed but the GN was not easily deterred. Nonetheless, the end of the CPR's monopoly to Phoenix was to come later rather than sooner. Various legal actions by local landowners and the CPR kept the Great Northern at bay and the line to Phoenix was not completed until March 1905. In the meantime, the CPR improved its trackage in the area and expanded its services, enjoying its monopoly while it could.

As for the Kettle Valley Lines, the railway struggled on, but it was not in a position either geographically or financially to challenge seriously either the CPR or the Great Northern, although through litigation and political action, it was not without means of influence and what the GN would have described as obstruction. The federal charter granted in May 1901 had given the railway greater powers including the right to build north up the North Fork of the Kettle River to the Franklin mining camp. Early in 1903, a federal subsidy was granted for this extension to Franklin Camp at a rate of $3,200 a mile (nearly $2,000 a km). This did not affect the Republic & Kettle River in Washington State, which in 1905 was renamed the Spokane & British Columbia

in an attempt to sound more important to investors and reflect its ambitions and the possibilities of its charter, but this long, costly, and unnecessary extension was never built. Trackage was extended up the North Fork of the Kettle and eventually reached Lynch Creek in 1907, just 20 miles (32 km) from Grand Forks. In 1919, after the railway was taken over by the CPR, it was extended to Archibald, a short distance to the north to serve a mine, although the line, then known as the North Fork Subdivision, was abandoned in 1935. To serve its small fleet of locomotives and other equipment, the Kettle Valley Lines built a roundhouse and repair shops on property known as the Ruckle Addition at Grand Forks. Its greatest significance, however, was in providing the CPR with a legal and corporate framework for expanding across the southern interior of British Columbia under the name of the Kettle Valley Railway. The Kettle Valley Lines' modest operation at Grand Forks and the extension north to Lynch Creek explains why, after 1911 when it was taken over by Canadian Pacific as the basis for the Kettle Valley Railway, the KVR technically had some trackage at Grand Forks, while most of its system was the then new railway built west of Midway to the Okanagan and the CPR's main line connections across the Fraser River from Hope and, via the CPR at Merritt, to Spences Bridge.

The little town of Curlew was an important stop for the passenger trains to Republic and it became a junction point once the tracks were built west to Midway and Molson. The photo above shows a northbound passenger train stopped beside the depot and water tank. It will soon steam out of town for Grand Forks and then swing south to Marcus and connections to Spokane.

–Jerome Peltier, Eastern Washington State Historical Society, L84-327.1357.

Obscure Narrow Gauge...

The Belcher Mountain Railway was a three-foot-gauge line built from ore bunkers in the mountains about eight miles (13 km) to Karamin (later Belcher Siding), Washington, on the Great Northern south of Curlew. Ben Scott, at far left in the photo above, was the engineer, and Fred Murphy, second from right, was the brakeman. The photo was taken about 1915. The engine is No. 2, a 2-8-0, probably from the Alberta Railway & Coal Co. The railway operated only between 1906 and 1915, but a trestle, below, survived into the 1970s. A report noted, "while the problems of transportation have been solved in a way, the facilities are unsatisfactory. The narrow-gauge railroad is difficult of operation, the grade is steep, and derailments have been of frequent occurrence."[13]

–Dave Wilkie Collection.

Excursions and Special Trains...

On the facing page: A Kettle Valley Lines excursion train pauses for the photographer on the railway north from Grand Forks to Lynch Creek. Everyone dressed in their best for the event, including the young ladies in their white dresses, but it would appear that there were still seats available in the two elderly open-platform combination baggage and coach passenger cars. One of these elderly cars is also in the photo above. The line to Lynch Creek and Archibald was abandoned in 1935.

–Dave Wilkie Collection.

A Great Northern special, decorated with flags and red, white and blue bunting, makes a festive sight behind 2-6-0 No. 480, former Spokane Falls & Northern No. 3, in 1908 on the Marcus-to-Republic line. Unfortunately, the exact location and the occasion were not recorded but such excursions were typically run on public holidays like the Fourth of July in Washington, and Victoria Day in May or Dominion Day (now called Canada Day) in British Columbia. There were enough bartenders in Grand Forks and vicinity to sponsor their own excursion train, pictured at left, on May 8, 1906. The engine is Spokane & British Columbia (Kettle Valley Lines) No. 1, an elderly 4-4-0. Apparently a good time was had by all, but memories likely were a bit vague on the details the next morning.

–Dave Wilkie Collection above; Boundary Museum, at left.

A Holiday Special...

"The Dominion Day [July 1] excursion to Grand Forks from Phoenix drew 302 passengers which required five boxcars with seats placed in them in addition to the three combination cars available."

–*The Phoenix Pioneer*, July 6, 1901.

Phoenix, the Mines and the Smelters

Record Production and Experienced Men Brought Phoenix to the Top...

"Monday the Granby mines again broke their 'one day' record for ore shipping. Christmas day these mines set out 60 carloads of ore, which with eight cars from the Snowshoe made 2,000 tons for the 24 hours. Monday the Granby mines alone sent out 73 cars or about 2,200 tons. As far as is known this is the largest tonnage ever sent out from one group of mines from any camp in the Province in a single day."

—*The Grand Forks News-Gazette*, JANUARY 10, 1903.

"Mr. Williams probably has more friends—of the kind that count—than any other mining man in the Boundary. It is not too much to say that everyone of the 400 men employed at the Granby Mines will regret to see him leave, to say nothing of hundreds of citizens of Phoenix and other parts of the Boundary. His uniform courtesy and consideration have won him a warm place in the hearts of the entire community."

—*The Phoenix Pioneer & Boundary Mining Journal*, JUNE 11, 1904.

William Yolen Williams was Granby's mines manager and superintendent from March 1898 to June 1904. He recruited fellow Welchmen to work at Phoenix and was very popular.
–*The Mining Record*, 1904.

THE CITY OF PHOENIX DEVELOPED QUICKLY around the mines four to five hundred feet (150 m) below the rounded peaks of the mountain ridge between the Granby River (then known as the North Fork of the Kettle River) on the east, and Boundary Creek on the west. Montezuma Ridge, Deadman Ridge and Spion Kop flanked the town to the north, and Knob Hill rose to the south. As miners, labourers, businessmen, families and others moved to the region to find work or establish a business in the seemingly never-ending boom, the city prospered. By 1900, already with a board of trade, its population was about 800 and it was formally incorporated as a city, giving it claim to the title of the highest incorporated city, at about 4,400 feet (1340 m), in Canada. Town lots sold quickly as property was placed on the market by G. W. Rumberger in the "lower townsite," and in the "New York" townsite on the hillside above. The Granby Company platted townsite lots and people were said to come from hundreds of miles away to buy corner lots for $600 or inside lots for $500, which was a substantial sum. In fact, *The Phoenix Pioneer and Boundary Mining Journal* reported that the revenue from the lots sold as parts of the surface rights to the claims was "sufficient to meet the initial expense of building the company's smelter at Grand Forks." In the summer of 1901 the Dominion Copper Company sold off surface rights to some of its properties, completing the general form of the city.[1]

As production increased so did the population, reaching over 2000 in the busy years before and during the First World War. Its civic attractions included an array of hotels (including the Brooklyn, said to cost $20,000 to build), saloons (reported to have been at least 17) and gambling halls with as many as 30 professional gamblers all after the miners' pay packets, a red-light district (as was common in frontier mining towns), a substantial range of churches of the various denominations (as an appropriate moral counterpoint to some of the previously mentioned attractions) which usually established themselves in the early days of the mining boom towns, a fire hall and city hall, a hospital, a post office, a school, a union hall and several ice rinks. Some of these amenities spoke of a more settled community, and there was even a large curling rink for which $10,000 was

Growing Output...

"Many mining men express the opinion that the output of the mines near Phoenix itself–mines such as those owned by the Miner-Graves group of companies–the Dominion Copper Co's mines, and the Snowshoe, will very soon exceed the output of the Rossland mines, and in doing so demonstrate to the world the extent and stability of British Columbia mining."
–*The Mining Record*, APRIL 1901, P. 121.

Photo facing page...
Phoenix in its heyday was a booming frontier mining town, intoxicated on the wealth of its ore and the promises of a limitless future.
–WILSON COLLECTION, OKANAGAN ARCHIVE TRUST.

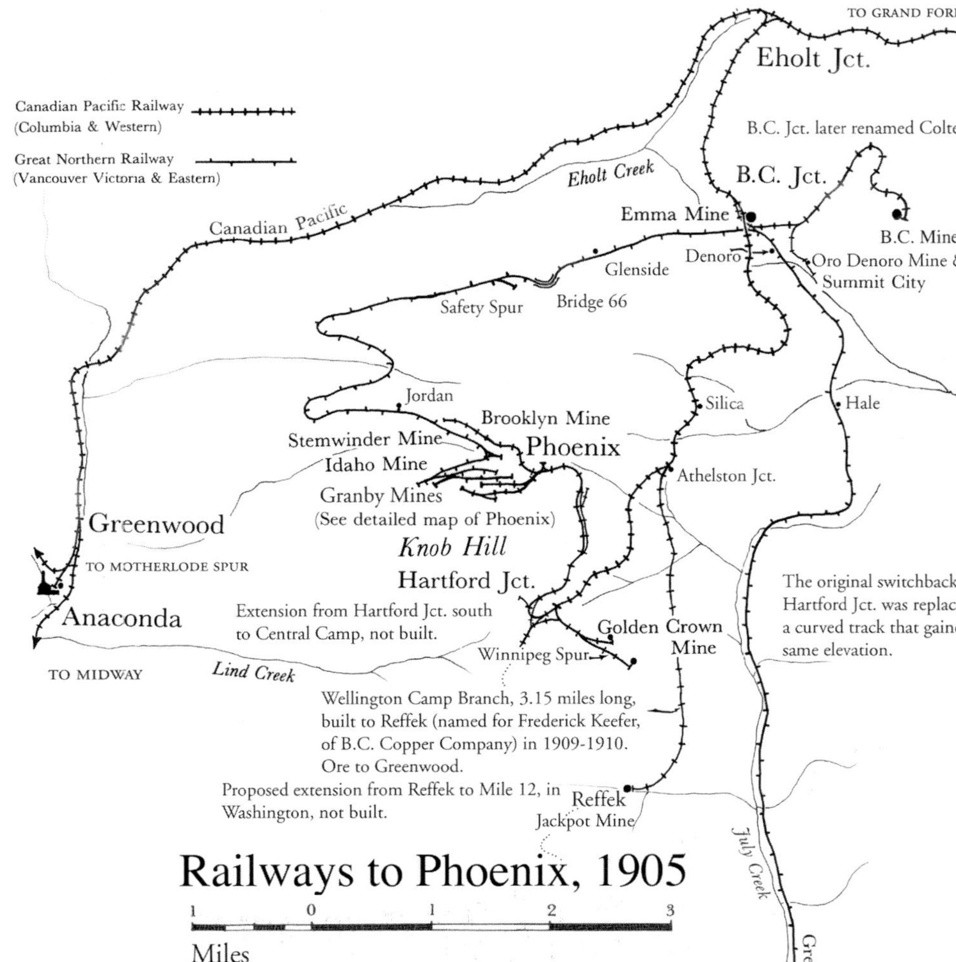

NORTHBOUND TRAINS INFERIOR DIRECTION. Second Class 849 Mixed / Daily Ex. Sun.	Miles from Eholt	Telegraph Stations	PHŒNIX SUBDIVISION STATIONS	Telegraph Calls	SOUTHBOUND TRAINS SUPERIOR DIRECTION. Second Class 850 Mixed a Daily Ex. Sun.
11.30	9.7	D	.... PHOENIX ....W	P	15.50
			2.2		
f 11.40	7.5		.... HARTFORD ....W		f 15.31
			2.0		
f 11.52	5 5		.... ATHELSTON .... Jct. with Reffek Sub'd		f 15.15
			0.6		
f 11.57	4.9		.... SILICA		f 15.09
			2.7		
f 12.15	2.2		.... COLTERN		f 14.47
			2.2		
12.30	.0	DN	.... EHOLT ....C W	KN	14.30
a Daily Ex. Sun. 849			PACIFIC TIME		/ Daily Ex. Sun. 850

Brakemen ride the wooden ore cars of a loaded train from Phoenix, about a mile from Eholt, on the way to Grand Forks. At left, a Canadian Pacific ore train crosses over the B.C. Mine spur at Mile 2.4 on the Phoenix Subdivision. High-grade ore frequently was carried in boxcars as is probably the case here. The employee timetable is from 1904. Timetables referred to the Phoenix line as the Phoenix Branch and later as the Phoenix Subdivision. Similar terms were applied to other mining branches and short branch lines were sometimes officially "Spurs." –Canadian Pacific Archives, 1616 and 1627.

The city of Phoenix, incorporated on August 31, 1900, grew rapidly from a rough mining camp into a prosperous community that proudly boasted of its amenities. By 1904 the population was over 1,000 and it continued to grow to at least double that figure. The city had a board of trade, a school board, churches of several denominations, a magistrate, several banks, a multitude of hotels, a requisite number of drinking establishments and breweries, and of course the mines. Societies included the Phoenix Miners' Union, No. 8, which met on Saturdays, the Knights of Pythias Phoenix Lodge, No. 8, the Fraternal Order of Eagles, and the A.F. & A.M, Masonic Lodge, No. 36. Such organizations provided support and assistance to members and an important part of the town's social life. –WM. NOTMAN & SON, NO. 3680, MCCORD MUSEUM.

From Mining Camp to Bustling City...

Phoenix Arose on Top of a Mountain of Copper Ore...

The face of Phoenix was almost constantly changing in its first decade as mines developed, large ore bunkers were built, new power plants and shaft housings were constructed, and the city itself grew to house and serve the workers and their families. This Notman photo from 1903 shows the town and some of the Canadian Pacific yards as the boom intensified. The facing page shows scenes from the next few years as the city expanded and many new businesses were established. –Wm. Notman & Son, No. 3682, McCord Museum.

The Victoria Shaft and the Old Ironsides dominated the skyline. Above left is the Victoria's big shaft building under construction in 1907. At centre left, a Davenport tank engine, pulls ore cars from the No. 3 Tunnel with the Old Ironsides behind. At left miners test their skills in a drilling contest.
–Hal Reiger Collection top; BC Archives centre, 18380; and Joe Tremblay Collection, courtesy Doug Cox.

Phoenix was situated in a basin between two ridges. Boardwalks and staircases began to knit the city together, avoiding the stumps and mud of the roughly hewn townsite. Above, the Granby Hotel is at far left and new bunkers and crusher for the Granby No. 3 tunnel are at right. The Old Ironsides ore bins are in the centre. The Granby Hotel, one of the largest in Phoenix and a residence of many miners, was on the south side of town overlooking the city. It was the venue for many a banquet and reception during the short life of the city. Less grand was the Hotel Phoenix, shown festively decorated in flags, bunting and greenery.
–BC Archives, 18372, above; Joe Tremblay collection, courtesy Doug Cox, left.

Granby's offices at Phoenix were strategically located within sight of the Old Ironsides and Granby mines and close to the Granby Hotel.
–B.C. Photo Engineering Photo, Robert Turner Collection.

Fourth of July Robbery...

"Burglars Celebrate the Fourth of July in Style. Excursion into the Hot Air safe nets them $750. About $600 of this belonged to the railway and the balance to Dominion Express Co.

"On this day arrangements had been made with the CPR and a special went out at 8:30 a.m. Conductor Ployart in charge with an excursion party bound for Republic. Agent A.B. Hall, after issuing the tickets and closing up the business office got on board and affirms that he carefully closed the safe and fixed the combination. A. McLennan of the CPR freight depot who was near Hall at the time vouches for the closing of the safe too. An iron crowbar and 14-pound hammer found in the office had been taken from the Boundary Iron Works near the depot."

–*The Grand Forks Gazette*, July 8, 1905.

subscribed for its construction by nearly 400 residents in just 10 days. Curling bonspiels were popular events in the mining towns in southern British Columbia and teams travelled back and forth, often on special trains, for the competitions.

For all the diversity and frontier atmosphere of Phoenix, it was claimed that one daytime and one night-time policeman, along with a magistrate, had no trouble keeping the peace. C. M. Campbell, who was resident manager of the Granby Company at Phoenix, wrote in 1921, as Phoenix was being abandoned, that the most exciting event in the town was when "'Tuck' Collins tossed an imitation stick of dynamite, with lighted fuse attached, into the crowded back room of the Brooklyn Hotel and yelled 'Fire'."[2]

Miners came to Phoenix from all across Canada and in particular from the Maritime provinces, as well as many parts of the United States. There were also significant numbers of miners from North Wales, encouraged by William Yolen Williams, until 1904 superintendent of the Granby mines.[3] Others came from England and Scotland as well as from Eastern Europe. Swedish miners comprised an important part of the population at Phoenix and there were also smaller numbers of Finnish and Italian miners. But this cosmopolitan atmosphere was not without limitations or prejudices. C. M. Campbell noted with considerable approval in his 1921 article that "early in its life the Miners' Union took a decided stand on the Chinese question and no Oriental ever resided in Phoenix."[4] Although management may have supported the union position on Asian labour in the mines, it showed little inclination to deal with the unions on other issues. As John Fahey notes in his biography of Jay P. Graves, "[Granby's] managers ignored the union." A strike for higher wages in 1907 ended on Christmas Day with the miners accepting lower pay and foregoing recognition of the Western Federation of Miners.[5]

The city of Phoenix effectively was divided into two parts by the Brooklyn, Stemwinder and Idaho mines with their large shaft structures and ore bins, waste dumps and glory holes, but the city hall and fire hall were strategically placed in the middle, right between the mines. Commercial and residential areas developed to the east and west. Two railways soon served the town and the large mine structures and waste dumps dominated the views from any street in the city. City streets usually were given the names of the major mines, companies or early claims. Avenues, running generally east and west, included Granby, Old Ironsides, Knob Hill, Aetna, Victoria, Dominion and Brooklyn. Streets, running north and south, were less commercially named in the eastern part of the city and included the appropriately designated School Street and Church Street, as well as pragmatically numbered First, Second, Third and Fourth streets. To the west, beyond the mines and dumps in the other half of the city, were Phoenix and Cimmeron streets, both of which led to the Great Northern depot, and Idaho and Banner streets. The Canadian Pacific station was located at the eastern end of the city at the end of Knob Hill Avenue below a small peak patriotically called Spion Kop, for the scene of the January 1901 battle during the South African, or Anglo-Boer, War. (Spion Kop, or "Lookout Hill" in Dutch, figured prominently in the intense and bloody fighting that led to a British defeat near Ladysmith in Natal.) The Boer War, as it was usually called in Canada at that time, raged from October 1899 until the end of May 1902, and was very much on the minds of the miners and investors in the copper camps of southern British Columbia. Many people from British Columbia had volunteered to join the Imperial forces fighting in South Africa. And the rich mining discoveries in South Africa had led to many fortunes, and bolstered the speculation and optimism that the mines of southern B.C. would be equally productive and profitable in a more politically stable region.

Although Eholt was located north of Phoenix, the Canadian Pacific's branch line, because of a twist of topography, actually entered Phoenix from the south. Its trackage made a loop south of the city at the Wellington Camp, and swung back into the depression where the city and the major mines were concentrated. Facilities included a yard with four main tracks, a wye for turning locomotives and other equipment, as well as the station. From there one spur led to the Old Ironsides Mine, about 700 feet (215 m) above the city, and a second, the Brooklyn Spur, went along the north side of town to the Stemwinder and Brooklyn mines.

Ironically, the Great Northern, which a few years later approached Phoenix from the south following July Creek upstream, entered the city from the northwest. The tracks made a long loop to the north of the city, crossing under the CPR at B.C. Junction, where the Emma Mine was located, and curving around to the south. The tracks came into the city over a large trestle that carried the tracks over westward-flowing Twin Creek that drained down towards Greenwood. The trestlework also supported part of the trackage for a wye, and from there supported the tracks leading to the depot, a small yard and the ore-loading bunkers of the Idaho Mine. A two-stall engine house was built to service the engines. The GN tracks continued south from the wye and over switchbacks to reach the lucrative Old Ironsides Mine and the Victoria Shaft on the south side of the city, at an elevation of about 4,600 feet (1400 m), or 600 feet (180 m) above the GN station. A large yard was built around these important loading facilities.

For all the development, investment and air of permanency that surrounded the mines and smelters, Phoenix and its railways and the smelters existed for the sole purpose of mining, processing and shipping ore, and it would be the value of the ores that determined their fates.

The Canadian Pacific's station at Phoenix was a modest but functional wooden-framed structure located on the northeast side of the city. It was typical of the stations built along the railway to the Boundary mines, and is shown above when nearly new about 1905, during a quiet interval, and at right about 1910, with Granby's Victoria Shaft and ore bins in the background, as a mixed train sits outside having recently arrived from Grand Forks.

–Boundary Museum 986.194.001, above; W. Gibson Kennedy Collection, right.

Phoenix in 1908 was a bustling city. This panorama, looking southeast, was taken from Spion Kop, above the CPR station, far left. The Victoria Shaft and ore bins are near the centre while below and to the right are the main business parts of the city. Other major mines were behind Monarch Hill in the background. –W. J. Carpenter, Greenwood Museum.

One of the CPR's mighty Shay engines rolls along the ridge high over the city of Phoenix. The Great Northern's wye, two legs of which are on trestles, the water tank and station are just below the locomotive in the middle distance. The construction of the GNR to Phoenix is described in subsequent chapter. –J. McRae Photo, Wilson Collection, Okanagan Archive Trust.

The detail view at left, enlarged from W. J. Carpenter's panorama, shows the upper town with the Granby Hotel and the Old Ironsides Mine behind it at right. The Granby, shown above, featured a large dining room and banquet facilities, and served as a home for miners working at the camp. The frontier mining camps had a high proportion of single men, many of whom boarded in the hotels.
–Joe Trembley Collection above, courtesy Doug Cox.

Eholt became a bustling junction on the Canadian Pacific between Grand Forks and Greenwood, alive with trains and activity. The bird's-eye view shows the busy rail yards and community. The roundhouse was accessed from the Phoenix branch. It took large crews of men to keep the steam locomotives running and this view at left shows the Eholt shop workers in front of two of their 2-8-0s. The scenes above show the eight-stall roundhouse, the complex yard tracks and three 2-8-0s and a caboose in front of the shops.

In the scene above, photographed between 1902 and 1905, a westbound passenger train is on the main line beside the Eholt station and water tank; an ore train behind engine No. 913, built in 1898, is about to enter the main line eastbound for Grand Forks and the Granby Smelter; and, on the tracks curving away to the left, the mixed train to Phoenix is preparing to leave the station for the climb to the mines. An empty ore train, powered by a 2-8-0 on the front and another one pushing, will follow the mixed to Phoenix. Steam and smoke linger in the damp winter air. By this time there were usually 15 CPR locomotives working on the Boundary lines.
–Opposite page, top, J. McRae Photo, Wilson Collection, Okanagan Archive Trust; Roundhouse views, Dave Wilkie Collection far left; Penticton Museum and Hal Reigger Collection, right hand column. Above, J. McRae photo, Dave Wilkie Collection.

Busy Eholt Thrived on the Traffic to Phoenix...

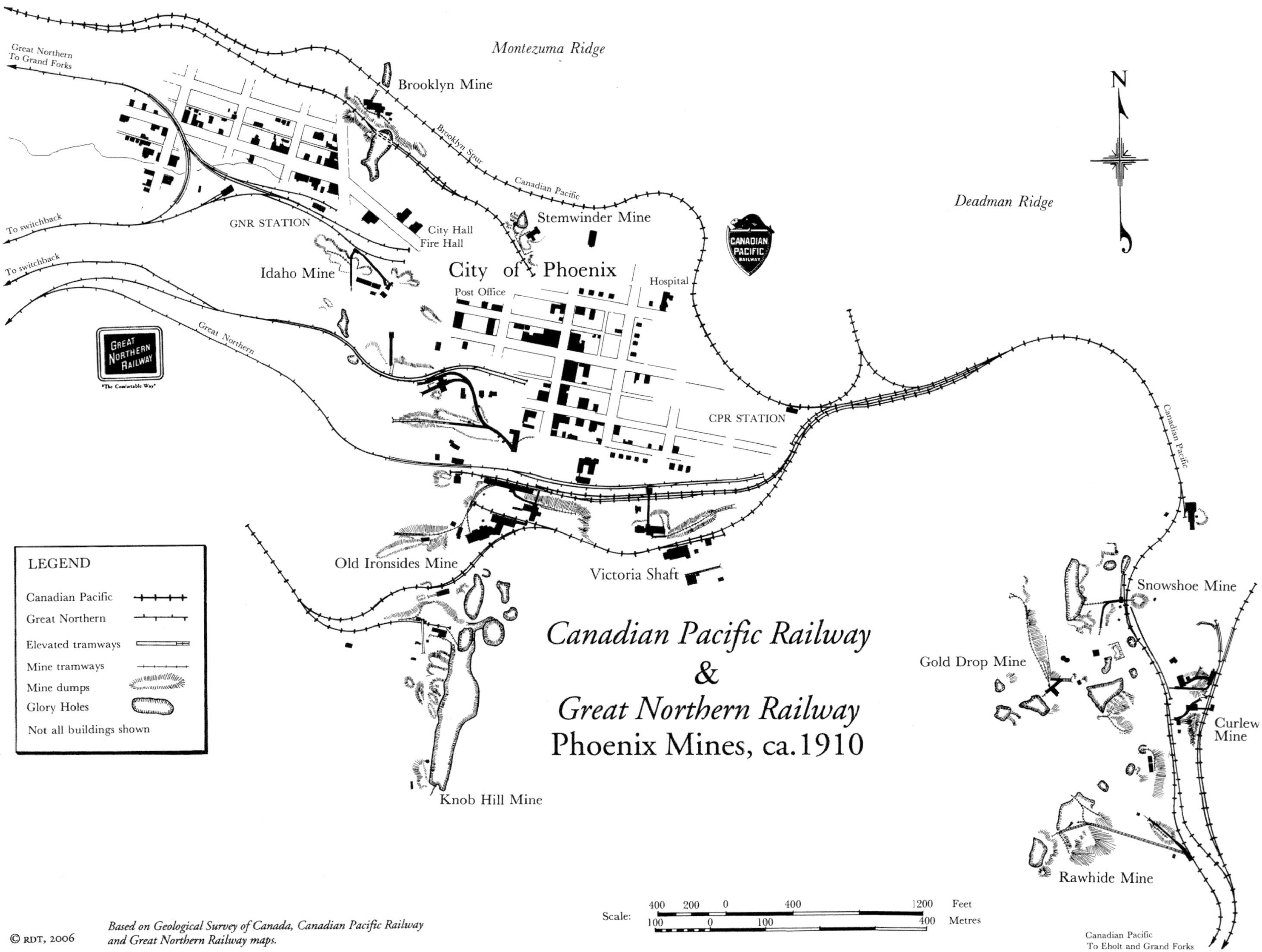

Great Northern
To Grand Forks
Montezuma Ridge
Brooklyn Mine
Brooklyn Spur
Canadian Pacific
Deadman Ridge
N
To switchback
GNR STATION
City Hall
Fire Hall
Stemwinder Mine
CANADIAN PACIFIC RAILWAY
To switchback
Idaho Mine
City of Phoenix
Hospital
Post Office
GREAT NORTHERN RAILWAY
"The Comfortable Way"
Great Northern
CPR STATION
Canadian Pacific
LEGEND
Canadian Pacific
Great Northern
Elevated tramways
Mine tramways
Mine dumps
Glory Holes
Not all buildings shown
Old Ironsides Mine
Victoria Shaft
Snowshoe Mine
Canadian Pacific Railway
&
Great Northern Railway
Phoenix Mines, ca.1910
Gold Drop Mine
Curlew Mine
Knob Hill Mine
Rawhide Mine
Scale:
400 200 0 400 1200 Feet
100 0 100 400 Metres
Canadian Pacific
To Eholt and Grand Forks
© RDT, 2006
Based on Geological Survey of Canada, Canadian Pacific Railway and Great Northern Railway maps.

The Phoenix and Mother Lode Mines

The Phoenix Ores…

"The ore throughout is remarkably uniform and is almost self fluxing. It consists of finely disseminated chalcopyrite, with pyrite and hematite…. The chalcopyrite carries all the copper, gold, and silver values, the average ore containing 1.2 to 1.6 percent of copper with about $1.00 in gold and silver to the ton."

—O. E. LeRoy, *The Geology and Ore Deposits of Phoenix,* 1912, p. 21.

At the Mother Lode Mine…

"A number of pillars and floors with a total height of 300 feet, on the east side of the old glory hole were dropped by the largest blast ever set off in mining history… 49,550 lb. [22 475 kg] of 40 percent powder was used to charge 4,834 holes from 14 to 15 feet [nearly 3 m] deep… 87,048 feet [26 532 m] of electric wire was used to [ignite] these charges. Between 400,000 and 450,000 tons of ore were broken down."

—Minister of Mines, *Annual Report,* 1913.

Supplying Ore for the Smelters...

"When D.C. [D'Alton Corry] Coleman was made Superintendent of the CPR Kootenay-Boundary lines... he decided... to keep one locomotive busy in and around Phoenix all the time gathering up the loaded ore dumps at the several mines and assembling them at Hartford Jct. Thence the trains will be sent direct to the smelters... without further switching."

—*The Phoenix Pioneer,* October 5, 1907.

The mining of the copper ores at Phoenix and the Mother Lode northwest of Greenwood impressed casual visitors and mining men alike. The facilities developed rapidly and the mines were modern, well equipped and productive. The scale of the operations was dramatic, and as exploration and more drilling went on, the ore continued to be plentiful but unfortunately of low copper and mineral concentration. Nonetheless, with railway development and the nearby smelters, it could be mined efficiently and economically.

The Phoenix ore body was substantial. "The west limb," noted O. E. Le Roy in his 1912 study of the district, "is 3,200 feet [975 m] long and 1,000 feet [300 m] wide; while the east limb is 2,250 feet [685 m] long and from 350 to 1,000 feet [100 to 300 m] wide. The curved connexion to the south is about 2,000 feet [600 m] long and from 200 to 700 feet [60 to 215 m] wide. The actual horizontal dimensions, however, are much greater, as a considerable area is overlain by… sedimentary and igneous rocks. The thickness of the west limb along an east to west axis varies from 160 to 350 feet [nearly 50 to 100 m]…. Along the north and south axis of the same limb the maximum thickness in the central portion is about 350 feet [100 m]." Much of the ore in this complex of deposits was near the surface, but substantial quantities lay below surface rock and overlying gangue or non-mineralized rock. The mines systematically worked at removing the ores from these deposits. By the end of 1909, 73,679 feet (22 457 m) of tunnels had been driven in the Phoenix mines and more were being opened at a rate of about 1,000 feet (300 m) a month.[1]

Mining engineers described three principal methods used to mine the copper ores in the Boundary District during the boom years at Phoenix and the Mother Lode deposits.[2]

Deep under Phoenix and through the surrounding ridges, miners dug into the copper ore bodies. The upper photo from 1908 shows a loading chute and mine car below a stope. The lower photo shows the heavy timbering, as big as the miner, sometimes needed in the mines. This photo is from 1919 as dismantling proceeded.

–Dave Wilkie collection.

Cross section of Granby mine workings through a stope.

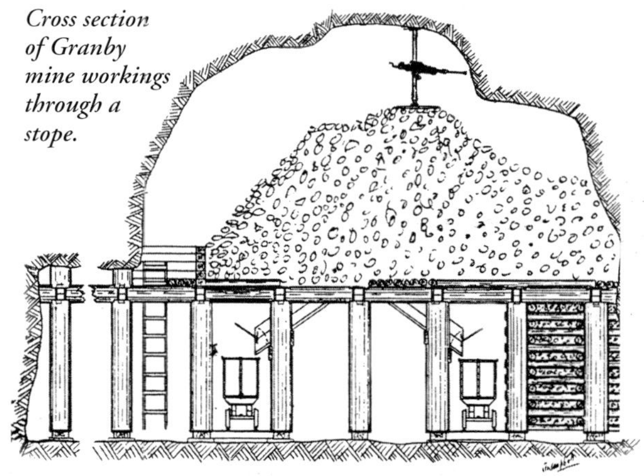

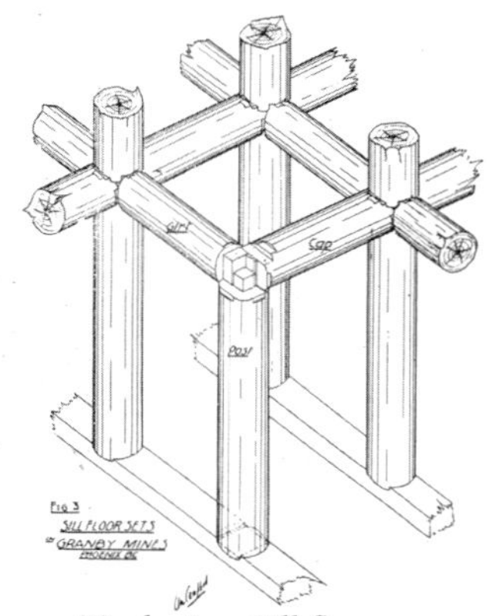

Timbering: Sill floor sets

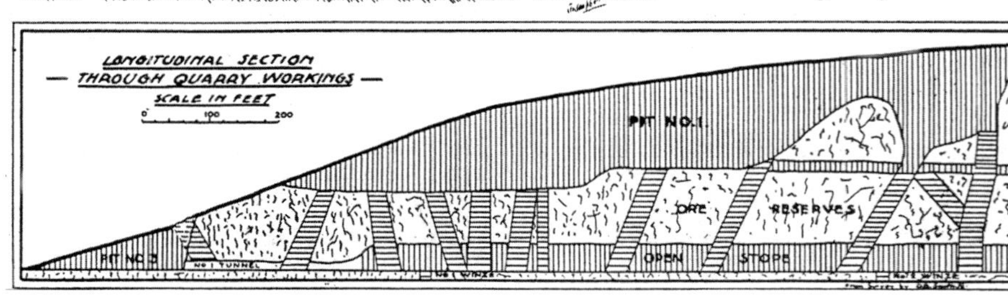

Open Pits, or Open Cast and Glory Holes

The ore was often in large pockets that could be mined without extensive and expensive timbering. Called "glory holes," these ore deposits were the delight of the mining engineers. The glory holes of Phoenix helped make the camp famous. The deposits of ore were close to the surface and the walls of the deposit converged gradually with the ore forming a large pocket. A drift, or tunnel, was run under the ore near a point where the top and bottom walls (the perpendicular wall and the foot wall) of the deposit met. Rises or short tunnels were run upward to the ore body and chutes were cut or constructed at the bottom. Drilling and blasting proceeded from the bottom, and the ore was removed from the drift. If the tunnels could be maintained at about 45-degree angles, then no shovelling of the ore was required. In an open pit, or open cast, the overburden was removed and the ore blasted and removed from the surface. When mined out, the glory holes were like huge chambers or caves in the surrounding rock. In some of the large deposits near the surface, steam shovels were used to remove the overburden and to load the ore after it was blasted from the solid formations. The big blast in 1913 at the Mother Lode Mine, referred to in the quotation at the beginning of this chapter and shown on page 87, illustrated the mining technique at an extreme. However, sometimes such excessive measures caused problems. In this case cleaning up after the blast took some time and actually disrupted the steady flow of ore to the smelter.[3]

Rock Pillars and Stopes

Underground mining of the Phoenix ores was usually possible without extensive and very expensive timbering of the workings. The use of stopes and rock pillars was similar to the extraction of ore from glory holes; however it was deeper underground. Where the dip or angle of the ore deposit was steep—for example, over 60 degrees at the Mother Lode Mine—several drifts were run under the ore and from these, rises were cut upward into the ore body at about 30-foot (10-m) intervals. Steel-lined chutes were fitted at the bottom and from these the ore was loaded into mine cars for movement to the surface. The ore was blasted out and removed from the bottom of the cavity, which expanded from each of the drifts until the adjacent ones met. Larger boulders or chunks of ore were broken up underground so that they did not clog the chutes. As the work progressed up into the ore body, the drifts were relocated to higher levels and pillars of ore were left to ensure the stability of the workings. Later some of the ore in the pillars was removed.

Timbering Sill Floors Under Stopes

This method was similar to the stoping techniques just described, but the drifts entering the ore body had to be timbered for safety. Ore above the heavily or partially timbered drifts was then mined from rises that were cut upward into the ore, which was then allowed to accumulate over the timbered drift and was removed from below. At the Emma Mine, the ore sometimes was in deposits 20 to 35 feet (6 to over 10 m) above the drifts, and the tunnels might be 10–12 feet (3–4 m) wide. By working upward into the ore body, the miners could load the ore without shovelling, making use of gravity, and saving significant time and money.

Development work increased at the Phoenix mines as revenues began to flow back to the investors and the strong world demand for copper continued. Deposits near the surface were worked with steam shovels and by the fall of 1903 three were in operation on the Granby properties. In addition, two small Davenport steam locomotives were acquired to replace mules and horses in hauling the mine cars along the surface tramways and in the larger tunnels. Production was reaching 2,000 tons a day and the older methods could not economically sustain these levels and the increasing production of ore needed as more furnaces came into production at the smelter.[4] In January 1904 an old 4-4-0 (Kettle Valley Lines' No. 1) was tried out briefly as a switcher at Phoenix but it was found to be too light for the work, and larger CPR engines were assigned to the job.

A mining area such as this that saw rapid expansion usually settled into a phase of longer term production after many of the smaller holdings were purchased and

Two of the 70-horsepower Baldwin electric locos, top, running on 500 volts DC, at the No. 3 tunnel of Granby's Phoenix mine. In the other photo, this small low-clearance Davenport tank engine, shown in 1908, burned coke to reduce smoke, and worked at the No. 2 tunnel, crusher and ore bins.

–Dave Wilkie Collection.

Facing page: **The glory hole workings opened up the mountain ridge like a piece of Swiss cheese. At right is a Thew steam shovel used in the surface workings. At lower left, one of the electric locos used at the 400-foot [120-m] level. Period diagrams, lower right, show timbering and how the ore was drilled, blasted and loaded from a stope at Phoenix.**

–Dave Wilkie Collection, two photos above; BC Archives, 18380, left; C. M. Campbell, *The Mining Record,* 1902.

Granby's Victoria Shaft and ore bins dominated the skyline of Phoenix. The large structure was opened in 1907 with ore shipments over both railways beginning that summer.
–B.C. Bureau of Mines, National Archives of Canada, C22980.

Photo, facing page...
A CPR train eases by the ore bin at the Golden Crown Mine near Hartford Junction, Mile 7.6, on the Phoenix branch about 1901.
–Canadian Pacific Archives, 1614.

consolidated into the operations of larger companies. These corporations or syndicates usually had the capital and resources to develop the infrastructure and, if necessary, survive periods of low prices for the metals being mined. In the Boundary District the consolidation of mining activities occurred by the early 1900s. Although the district was blanketed with mineral claims, some staked with serious hopes of ore, others simply as speculative ventures, key properties soon emerged and became famous for their ores and the return on investment everyone hoped they would produce over many years. By the early 1900s, there were three major holders of mining properties in the district and they all built or controlled smelters. (In addition, there was the CPR's growing smelter complex at Trail, on the Columbia near the mines at Rossland.) These were the Granby Consolidated Mining, Smelting & Power Company, which built a smelter at Grand Forks, the British Columbia Copper Company, which built an equally impressive facility at Greenwood and soon also controlled the less successful Boundary Falls smelter of the Dominion Copper Company, and finally the Consolidated Mining & Smelting Company, a subsidiary of the Canadian Pacific, which also operated the smelter at Trail. A summary of the more important mines follows, and the three Boundary District copper smelters are described in the next chapter.

317

Thew Automatic Co.'s Standard Type No. 1 Single Truck Steam Shovel.

Steam shovels dramatically increased production in the glory holes and open pits at the Granby mines. The Thew Automatic Co.'s shovel was used effectively for loading ore and overburden.

–Robert Turner Collection, from *B.C. Mining Record.*

Third Steam Shovel ...

"Granby expects a third shovel to arrive from Lorraine, Ohio, later this month. This... shovel was bought from the CPR. Haulage of ore cars by horse or mule power will be replaced shortly by two small steam locomotives. They are from the Davenport Machine Works... These engines will have 9 inch diameter cylinders and 14 inch stroke. Fourteen tons in service. They should haul from five to seven mine cars each and will be used in the no. 1 and no. 2 tunnels of the mine, which have grown to some lengths, in late years. In view of the increased shipments to 2,000 tons per day... the old method was too time consuming."

–*The Grand Forks News-Gazette,* November 14, 1903.

Granby Consolidated Mining, Smelting & Power Company's Mines

The Grand Forks smelter, along with the Old Ironsides and Knob Hill mines controlled by the Miner-Graves Syndicate, and the Grey Eagle Mining Company were consolidated in January 1901 to form the Granby Consolidated Mining, Smelting & Power Company, with a capital of $15,000,000. Directors were Stephen Henderson; Campbell Miner of Granby, Quebec; J. H. McKechnie of Montreal; W. H. Robinson; Alfred Cornelius Flumerfelt of Grand Forks; Fayette Brown of Montreal; and two of the original entrepreneurs, Aubrey Lee White and Jay P. Graves, both of Spokane. Although a few additional mines such as the Monarch, Gold Drop and Curlew were purchased, the main output continued to come from the Old Ironsides and Knob Hill mines.

By 1907 Granby's holdings included 1100 acres (445 ha) of mineral claims around Phoenix, and 61 acres (25 ha) containing a limestone quarry near Grand Forks. As well, Granby owned 100 town lots in Phoenix and Grand Forks, and the smelter at Grand Forks. Brief descriptions of some of the principal mines follow.

Knob Hill – Ironsides Mine

Granby's original claims were the Knob Hill–Ironsides group that included the Phoenix, Fourth of July, Old Ironsides, Knob Hill, Victoria and others, equally colourfully named. Purchases added many additional adjacent and nearby claims, encompassing a substantial ore body that was the foundation for the company's extensive mining and smelting development in the Boundary District. Ore from this mine was extracted from glory holes, and by shafts, levels and stopes. By 1907, surface work had exposed the ore body for a distance of 3,400 feet (1035 m) with an average width of about 200 feet (60 m), and diamond drilling was being carried out to define the further extent of the ore body.

Rail haulage was important in the extraction process. At the upper workings and glory holes, a three-foot-gauge system used a 14-ton Davenport saddletank steam locomotive to haul steel and wooden ore cars through the mine's No. 2 tunnel located 250 feet (75 m) below the top of the ridge, and containing 56 ore chutes. Ore was crushed and then dumped into "capacious" storage bins before being taken by rail to Grand Forks. One hundred feet (30 m) below was the No. 3 tunnel that also had a railway system, but this one was electrically powered using a 70-horsepower mine "motor" or locomotive. The ore from this tunnel also was crushed before going to the smelter. Together the two tunnels could produce 2,000 tons of ore a day. The ore from the Nos. 3 and 4 tunnels was mined using stopes and loaded through chutes. The electric locomotives worked over a 4,000-foot (1220-m) three-foot-gauge system, using specially designed five-ton ore cars that carried the ore to storage areas near the shaft for haulage to the surface.[5]

The company's Victoria Shaft, along the ridge south of the city, had imposing surface structures. The Victoria Shaft itself had three compartments, two of which were used for ore, the third for the miners. The five-ton skips (small railway dump cars) used to move the ore could be hauled up the shaft at 700 feet (nearly 215 m) per minute, giving an impressive capacity of 1,000 tons per eight-hour shift. Working at full production, the mine employed 168 machine miners underground and another 112 men in surface trades. At this time the Gold Drop, which had originally been owned by Montreal interests and was later taken over by Granby, was also being developed. Canadian Pacific served the Victoria Shaft, No. 2 tunnel and the Gold Drop while Great Northern, after 1905, also provided rail connections to the Victoria Shaft and the No. 2 tunnel, and also reached the No. 3 tunnel along a switchback route into the valley below, where additional loading facilities were located (shown on page 122).

British Columbia Copper Company's Mines

The British Columbia Copper Company's holdings included the Mother Lode, a highly productive property, near Greenwood, and, through majority ownership in the Dominion Copper Company, large properties at Phoenix. The company also owned the Lone Star and Napoleon mines in Washington.

Mother Lode Mine

This large and productive mine, pictured on the following pages, was located on a group of claims of about 60 acres (25 ha) and several other properties. Situated just four miles (6.5 km) northwest of Greenwood beyond the mining camp of Deadwood, the mine was reached by a steeply graded Canadian Pacific spur line from Greenwood that connected the mine to the B.C. Copper Company's smelter and the CPR's line to Grand Forks and the east.

The Mother Lode was mined by glory holes and by the use of rock pillars and stopes. The ore body was extensive and the mine structures were well developed with electric power coming from Bonnington Falls. By about 1910, the mine's main shaft went down 475 feet (145 m) and had four compartments each four feet, six inches by five feet (1.4 x 1.5 m), with two used for removing the ore and waste, one for the "man cage" and one for ladders and pipe. Drifts and stopes reached a depth of 600 feet (180 m). When in full production of 1,200 tons of ore daily, the mine employed 137 men underground and another 39 on the surface, as well as office and clerical workers. Twelve horses worked underground to move the ore cars.

Emma Mine

Located at the northern edge of the major deposits and closer to Eholt than to Phoenix where the Great Northern crossed under the CPR, the Emma Mine was developed to extract copper ore (magnetite) that ran in a nearly vertical vein extending at least 250 feet (more than 75 m) below the surface. This ore, at 1.44 percent copper, was high in iron at 45.6 percent with 13.25 percent silicon dioxide and 11.2 percent calcium oxide. Gold averaged 0.307 ounces (8.703 g) a ton and silver was less at 0.027 ounces (0.765 g) a ton. The Emma Mine was jointly owned by the B.C. Copper Company and the Hall Mining & Smelting Company, with B.C. Copper holding three quarters of the shares. The mine was closed for a time, but reopened in 1916 and shipped ore to the Consolidated Mining & Smelting Company's smelter at Trail and came under CM&S control.

Oro Denoro Mine

The Oro Denoro, situated just south of the Emma Mine, lacked the clearly defined ore body of the Emma and was worked with glory holes and some tunnelling or "drifting," as the tunnels were called. The ore, different from the Emma's, contained about 1.5 percent copper, 31.3 percent silicon dioxide, 18.4 percent iron and 29.3 percent calcium oxide, but without significant traces of gold or silver. After mining, the ore was crushed and then loaded into a storage bin pending shipment to the smelter.

Nearer to Phoenix, the company controlled the Brooklyn, Idaho, Standard, Stemwinder, Montezuma and Rawhide claims covering an area of 177 acres (72 ha). The Brooklyn and Stemwinder mines were located along the northern side of Phoenix, and the mining operations almost divided the city into two sections. Situated along the lower slopes of Montezuma Ridge, the two mines overlooked the city and were served by the Brooklyn Spur of the Canadian Pacific. The Idaho was just a few hundred yards to the south but was closer to the location of the Great Northern's yard and station. More removed from the city was the Rawhide, located south of town on the CPR, near the Granby Company's Gold Drop and Consolidated Mining & Smelting's Snowshoe mines. Production at the Rawhide grew in the years before the First World War and climbed from 50,400 tons in 1911 to 267,347 tons in 1912, dropping slightly to 238,455 tons—about 600 tons a day—in 1913. As investment grew in smelting capacity and other facilities, the need for a steady flow of ore became critical and companies benefited from having several productive properties to assure reliable supplies of ore, and ideally a mix of ores that facilitated smelting and could contribute to the overall efficiency of the smelting processes.

Consolidated Mining & Smelting Company's Mines

This CPR-owned company operated the smelter at Trail and was later known as Cominco. The Snowshoe group was leased in 1906 and then a group of 11 claims

In a huge blast that broke down nearly half a million tons of ore at the Mother Lode Mine in August 1913, 4,834 holes, each 14 to 15 feet deep, were filled with a total of 49,550 pounds of 40 percent TNT. Over 87,000 feet of electric wire was used to connect the charges. It took many days to clean up the debris and get back into full production. –BC Archives.

Workforce at the Mother Lode Mine, 1910...

Underground workers

75 machine miners
6 timbermen
50 trimmers, mockers etc.
6 shift bosses

Surface workers

4 hoist engineers
3 compressor engineers
5 blacksmiths, with 5 helpers
4 carpenters
2 machinists, with 2 helpers
3 crusher men
3 firemen
1 electrician, with 1 helper
6 labourers

–Mines Branch, 1907–1908, *Report on the Mining and Metallurgical Industries of Canada*, p. 147.

The Mother Lode Mine, Shays, Bridges and the Sunset Mine...

Canadian Pacific's Shay 1902, with engineer Bob Colley in the cab and Ed Funk, who later became a conductor, standing next to the smokebox, was photographed at Phoenix soon after its delivery in 1902. The CPR Shays worked also the branch line to the Mother Lode Mine as these photos show. At right is the Sunset amidst a maze of trestles, bunkers and waste dumps.

–BC Archives, 39429, right; Robert Turner Collection far right.

William J. Carpenter of Spokane took this and other panorama views of the mines and smelters in the Boundary District. Some of his prints were huge, measuring nearly 2 x 6 feet (0.5 x 1.8 m) printed from a single negative in amazing detail. This photo captures the Mother Lode and Sunset mines near Greenwood on the Mother Lode Spur, as the short branch was called officially, in their peak years. The detail view shows one of the CPR's Shay locomotives at the bunkers of the Sunset Mine. The Shays are described on pages 110-111.
–Greenwood Museum.

The Mother Lode Branch...

The shaft house and other buildings at the Mother Lode Mine from the panorama shown on the previous pages.
–W. J. Carpenter photo (detail), Greenwood Museum.

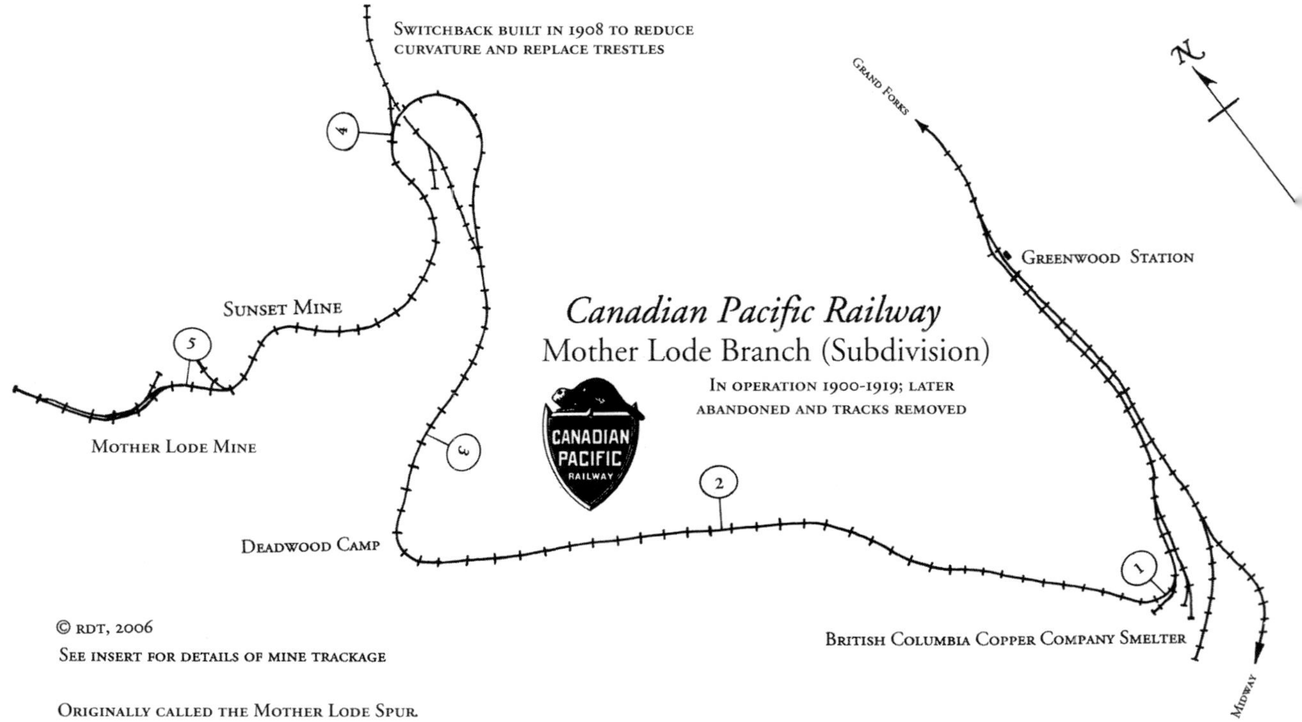

© rdt, 2006
See insert for details of mine trackage

Originally called the Mother Lode Spur.
Often spelled Motherlode.
Maximum grade: 3.8 percent.

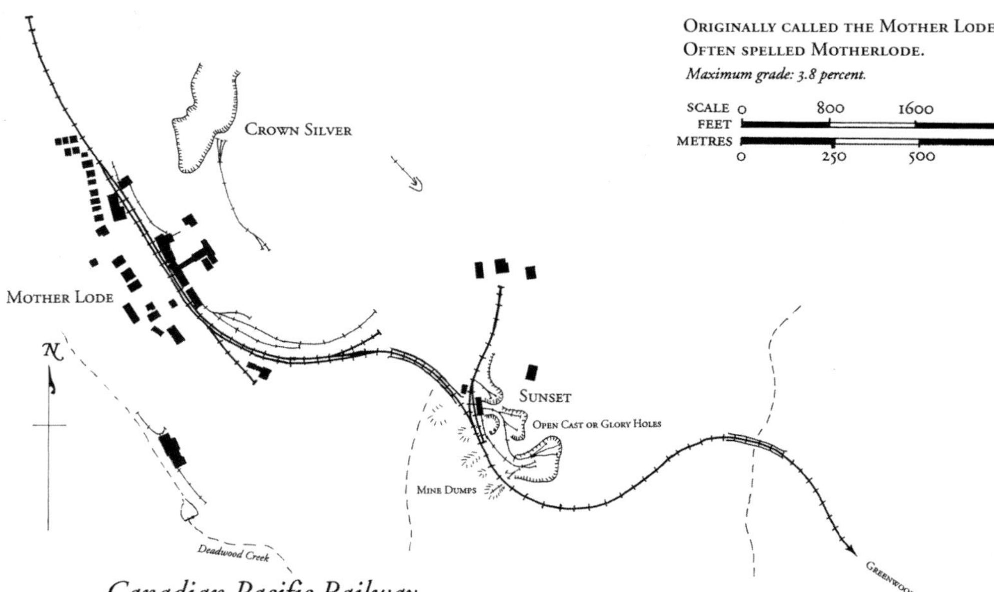

Canadian Pacific Railway
Mother Lode Branch (Deadwood Spur)
Details of Mother Lode Mine and Sunset Mine Trackage

Based on Geological Survey of Canada, Canadian Pacific Railway and Great Northern Railway maps.

© rdt, 2006

A Shay-powered ore train from the Mother Lode branch rounds the curve above the Greenwood Smelter about 1901. The branch was opened to traffic in mid-April 1900.
–BC Archives, 9994 (detail).

known as the Phoenix Amalgamated, which included the War Eagle claim, was purchased to expand the company's holdings in the area. The Snowshoe had changed hands several times before being leased. First staked in 1891 by G. W. Rumberger, the Snowshoe, which later proved to be very valuable, had been allowed to lapse, but was taken up in 1893 by two partners named Dengler and Douglas. They then bonded it to P. Clarke of Spokane in 1897 and two years later it came under the control of the British Columbia, Rossland, & Slocan Syndicate of London, which carried out some initial development work. This corporate name, including two of the most famous mining regions in the province, was undoubtedly chosen to appeal to investors in Britain. In 1901 the claim passed to the Snowshoe Gold & Copper Company of London. Large-scale production at the mine used a combination of glory holes as well as tunnels and shafts to reach the extensive ore body.

These claims were less than one mile southeast of Phoenix along the CPR branch line and were served by railway spurs to the mines, and they became important producers in the district. By mid-1910, the Snowshoe had yielded over 520,000 tons of ore with an estimated value of $2,013,361.[6]

The Emma Mine came under the control of the Consolidated Mining & Smelting Company during the First World War era and was reactivated. Ore was shipped over the CPR to Trail for smelting. Two carloads a day of ore, described as "a good fluxing ore," were being shipped in the winter of 1916, with 50 men employed in the mine.[7] But like the other Phoenix mines, it too closed at the end of the First World War when the reserves were depleted and the market declined.

Shay 1903 was the last of the three big engines acquired by Canadian Pacific for the mining branches. To increase its coal capacity, the sides of the coal bunker were extended.
–W. G. Kennedy Collection.

A Canadian Pacific Phoenix-bound freight storms onto the trestle spanning the Great Northern main line and interchange at Denoro. This is the Emma Mine, 2.4 miles (3.9 km) from Eholt and one of the important sources of traffic for the railways. The CPR had a spur to the B.C. Mine and it passed below the trestle from left to right. This classsic mining and railway scene is one of the few action photos of the railways around Phoenix from an era of slow-speed films and cameras with slow shutters. The mine structures show how much timbering and cribbing could be used around even a small mine. The locomotive is one of the CPR's L5a 2-8-0s, No. 1383, which in 1912 became the 3183. Like some other Phoenix engines, it ended its days with the CPR on Vancouver Island in the 1930s hauling log and coal trains.
–BC Archives, c-05264.

378—GRANBY COMPANY'S SMELTER, GRAND FORKS, B.C.
ENT. ACC. TO ACT OF PARL., CANA
YEAR 1903, BY WM. NOTMAN &
AT THE DEPT. OF AGRICULTURE.

The Boundary District Smelters

The New Smelter at Grand Forks...

"The Granby Smelter, ...now nearing completion, was located on its present site, owing to the exceptional water power in the vicinity. To utilize this power a dam 175 feet [53 m] across the top, 75 feet [23 m] from heel to toe at the bottom, and constructed of sawn timbers 12x12 inches [30x30 cm] and filled in rock was thrown across the North Fork of the Kettle River..."

—*The Mining Record*, June 1900, p. 207.

Now Four Furnaces...

"The Granby Smelter is now running all four furnaces and is making a record as the greatest smelter in the Dominion, if not the entire continent. Large new ore bins have been added to accommodate the heavy daily shipments which will be necessary in order to accumulate the ore reserve over... the 1,500 odd tons run through the plant daily...."

—*The Grand Forks Miner-Gazette*, April 5, 1902.

Dumping the Slag...

"The blast furnace slag [at Greenwood] overflows from the settlers into slag cars. The slag trucks used are oval trucks of the type of cinder car used at iron and steel works; each truck has a capacity of 250 cubic feet [7 m^3] of molten slag or about 25 tons. The slag trucks are hauled, one at a time, to the slag dump by 35-ton electric locomotives. Each slag truck is provided with a 15 H.P. motor, mounted on its frame, which furnishes the power to tilt the slag pot on the car frame, in order to discharge the molten slag. The tilting motors are operated from a controller located in the locomotive cab..."

—Alfred W. G. Wilson, *The Copper Smelting Industries of Canada*, June 1913, p. 131.

The ore typically extracted from the Boundary mines contained a complex mixture of metals and rock materials that had to be separated and treated to produce copper in the nearly pure form needed for industrial applications. The first step was usually the crushing and separation of the ore from obvious impurities at or near the mine site to reduce as much as possible the quantity of waste material. This lessened transportation and handling costs significantly.

The Phoenix ore was usually a chalcopyrite that contained about 1.2 to 1.6 percent copper along with small but very important traces of gold and silver. From this ore, Granby and the other local smelters produced the semi-pure form of copper called matte. The matte from the Granby Smelter, for example, contained 35–40 percent pure copper. Further processing through converters produced "blister copper" (also a form of chalcopyrite), 99.5 to 99.6 percent pure, and contained 4–6.5 ounces (113–184 g) of gold and 25–37 ounces (709–1048 g) of silver per ton. The term blister copper came from the appearance of the material, which was marked by bubbles in the surface of the copper from the sulphur dioxide gases formed and driven off in the smelting process. Granby cast this blister copper into 220-pound (100-kg) bars for shipment. During the smelting 65 percent of the sulphur in the ore was burned off (accounting for the acrid fumes downwind of the smelter that damaged or destroyed surrounding vegetation), producing slag that was composed of 45 percent silicon dioxide, 15 percent iron oxide, 22 percent calcium oxide, 3.8 percent magnesium oxide, 7 percent aluminum oxide, additional waste materials, and just 0.22 percent copper. Often the slag was reprocessed for further recovery. As well, smelter and furnace exhaust gases were screened to recover material suitable for making briquettes that were then burned in the furnaces.

The heart of the smelter was the blast furnace, which

Frederick Keefer was general manager of the British Columbia Copper Company's Greenwood smelter.
–*The Mining Record.*

Working at the Granby Smelter...

"The men employed are both union and non-union, and nearly every nationality of the white races is represented. The furnace and converter departments have a general foreman with shift bosses under him; each of the other departments has its own foreman. Shifts are eight hours each; wages vary from $2.50 per day for labourers to $3.50 per day for shift bosses."

–Alfred W. G. Wilson, *The Copper Smelting Industries of Canada*, 1913, p. 111.

Facing page photo...
After only a few years in operation, as this 1903 photo suggests, the Grand Forks smelter had become a substantial facility with a growing slag dump to mark its success.
–Wm. Notman & Son, No. 3678, McCord Museum.

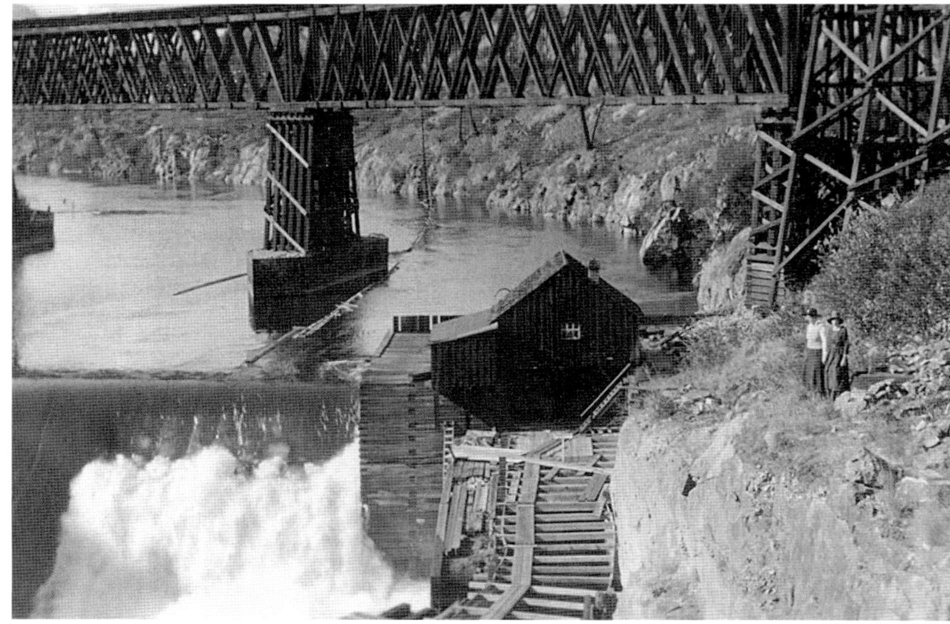

heated the copper and separated it from the other constituents of the ore. This was accomplished by heating the ore to a temperature above its normal melting point. Some impurities were burned off and the metal, which was heavier than the waste or slag, flowed to the bottom of the furnace and could then be poured off, as could the separated molten slag. The most common type of blast furnace used in the Boundary District was the water-jacket type, which was developed in the late 1800s. The water-jacket served to cool the outer parts of the furnace. In these furnaces air was forced through pipes with nozzles (called tuyères) into the furnace, creating the "blast." The blast was low in pressure but high in volume, increased the supply of oxygen to the burning coke in the ore and coke mix, and raised the temperature of the fire to 800 to 1000 degrees Fahrenheit or 500 degrees Celsius, sufficient to melt the ores, combust the sulphur and iron, oxidate the impurities, and separate the molten metals from the molten waste rock and minerals.[1]

Some ores required fluxing with limestone to reduce their melting point and make smelting practical, but the majority from the Phoenix and Mother Lode mines had sufficient fluxing properties that this was not required. Sometimes by combining different ores, a satisfactory mix could be derived that did not require added fluxing. The smelters sometimes competed with each other for ore, but they also had to establish processes tailored to the chemical characteristics of the different ore bodies. Some ores were easier to smelt than others, and smelters could not effectively deal with all types of ores in one facility.

Refining, usually by an electrolytic process, was not done in the Boundary District. This process yielded the 99.99 percent nearly pure copper needed for industrial applications, and recovered the very valuable traces of gold and silver that could determine overall profitability. Granby shipped most of its blister copper to the refineries of the Nichols Chemical Company in Brooklyn, New York, for final processing, and smaller quantities were shipped to Liverpool in England for sale in Europe. The connections with the operations of copper magnate William H. Nichol eventually led to his takeover of Granby in 1904.[2]

The mines and smelters required large amounts of power to drive the machinery in the plants and to run the compressors, fans and hoists used at the mine workings. At the smelters, a forced airflow was central to the operation of the blast furnaces. Electric power for the Boundary smelters and many of the Boundary mines usually came from hydroelectric generating stations. The Granby Smelter had its own power supply from a hydro dam located about one mile (1.5 km) from the smelter, and a powerhouse built on the North Fork of the Kettle River. The new lake, named, not surprisingly, Smelter Lake, flooded the valley north of the smelter. Additional power was purchased from the West Kootenay Power & Light Company's Bonnington Falls generating station on the Kootenay River just downstream from Nelson, B.C. It was transmitted over two 3-phase, 60-cycle, 60,000-volt power lines, and then it was stepped down to 440 volts to run the electric motors at the smelter. Bonnington Falls also supplied the Greenwood smelter with power, using over 85 miles (137 km) of transmission lines. Transformers reduced the 60,000-volt power to 550 volts used to run mine crushers and finally down to 110 volts for lighting. Many of the mines at Phoenix also obtained electric power from Bonnington Falls, photos of the city and mines clearly showing the many power lines.

The mining and smelting of the copper ores were complex processes and equally complex economics. When copper prices were low in 1913, the cost of producing ore from the Mother Lode was higher than the selling price of the copper; "however," the Department of Mines reported, "the excellent self-fluxing quality of these ores made it possible to buy and smelt a large tonnage of custom ore [from other mining operations] at a profit." By mixing the Mother Lode ores with the custom ores, the overall mix could be smelted at a low enough cost to be profitable. That year, the B.C. Copper Company treated 353,422 tons of Mother Lode ore at its smelter and 259,485 tons of custom ore, yielding 8,296,902 pounds (3 763 411 kg) of copper as well as gold and silver for earnings of $1,904,694.52.[3] Two large blast furnaces ran all year.

Photos, facing page and above...
The Grand Forks smelter expanded rapidly in the early 1900s as demand for copper grew and production at the Phoenix mines increased. The photos at far left show the smelter in its first years. The imposing stack was 152 feet (46 m) high. The photo, upper right, shows the power dam and railway bridge over the North Forks of the Kettle (Granby) River, before the GN built its bridge just upstream. Lower right is one of the blast furnaces.

The flume from the dam ran one mile from Smelter Lake to the Granby powerhouse, shown above. The dam itself was 175 feet (53 m) across and 30 feet (9 m) high, and the flume's dimensions were 9 x 12 feet (2.8 x 3.6 m). A further drop of 15 feet (4.6 m) from the dam to the powerhouse provided an output of 1,200 horsepower. Nelson engineer B. C. Riblet designed the power plant.

–Facing page, far left, Canadian Pacific Archives 1602 and below, 1637; at right, Boundary Museum, 991.023.002 and 991.055.243. This page, B.C. Photo Engineering Co., Robert Turner Collection.

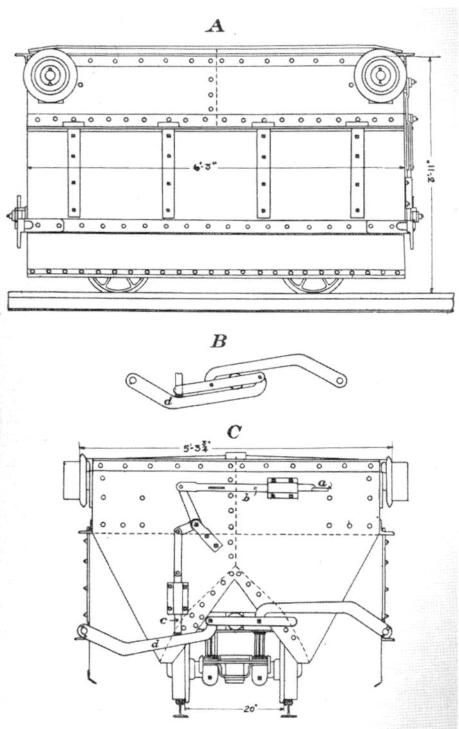

Charging cars from the Granby Smelter at Grand Forks.

–Alfred W. G. Wilson, *The Copper Smelting Industries of Canada,* 1913.

The Grand Forks Smelter: Granby Consolidated Mining, Smelting & Power Company, Ltd.

The massive Granby Smelter at Grand Forks dominated the economy of the city and the lives of most of the town's people during the early 1900s. When the smelter was "blown in" on August 21, 1900, the Grand Forks smelter had two double-decked, steel-jacketed furnaces measuring 160 x 44 inches (4.1 x 1.1 m). Initially, the company anticipated that 150 to 250 tons of ore a day, depending somewhat on the quality of the ore and its character, would be smelted, although the facility was designed with a maximum capacity of 500 tons a day. The main smelter building was an impressive structure measuring 64 x 70 feet (19.5 x 21.3 m), and was surrounded on three sides by substantial ore bins.[4]

Expansion followed rapidly and two more furnaces were added in 1901, raising the daily capacity of the plant to 1,500 tons of ore. Soon the smelter was enlarged still further, and then expanded again so that within a few years the smelter was described as the largest copper smelter in the British Empire, and the second largest in the world. With eight furnaces in use it could handle 3,400 tons of ore a day.

Slag from the smelter presented a major disposal problem. For the first three years of operation it was simply dumped in granular form, but it took up too much room in the restricted valley bottom where the smelter was located, and it also fell into the North Fork of the Kettle River (Granby River). Subsequently it was taken in molten form using small rail cars with six-ton pots and was hauled and dumped while still fluid, so that it hardened on site. By 1912, because a dump over a mile long and from 100 to 500 feet (30 to 150 m) in width had been created, the company returned to dumping the granular slag. Much of this massive, sterile industrial landmark would survive into the 21st century many decades after the closing of the smelter.

Production at the Granby Smelter was impressive. By mid-1912 the smelter had produced 7,952,939 pounds (3 218 554 kg) of copper as well as significant quantities of silver and gold. The increasing efficiency of the smelter was evident in the costs of processing the ore. In 1901 it was $4.77 per ton, but by 1912 it had dropped to $2.47. Moreover, the tonnage of ore shipped had grown from 172,258 dry tons of ore in 1901 to 723,024 tons in 1912. Reflecting market demands and production variations, a peak of 1,178,853 tons was shipped in 1908.[5]

The railways were very much a part of the smelter operations. With trains coming in from Phoenix and other mines over both the CPR and the Great Northern, the yards were busy. A switch engine was maintained by the two railways to serve the rail yards that were connected by crossovers.[6] Cars of ore were weighed on track scales, two being maintained at the smelter, with one for 36-foot (11-m) cars, and one for 50-foot (15-m) cars. Arriving ore cars were moved to the top of a terrace east of the main smelter where ore storage bins were located. These massive structures, built of timbers, could hold 12,000 tons of ore and were 756 feet (230 m) long with a height of 20–25 feet (6–8 m). These large bins made it possible to maintain an inventory of ores at the smelter to facilitate different mixtures as well as giving the smelter a stockpile in case of disruptions in production at the mines or interruptions in deliveries on the railway. In addition, the sampling mill had storage for 5,300 tons of ore. Ore was moved from the bins to the blast furnaces by the plant's narrow gauge railway.

Not only did the railways supply the smelters with ore and carry the copper matte to refineries, they also delivered large shipments of coke and coal as well as clay and sometimes limestone used in the smelting. Coal and coke were brought in from the Crowsnest mines at Fernie and Michel. The coke was transported in special rack cars or in boxcars with four long slots in the bottom for dumping. Clay, used in lining the converters to produce the blister copper, came from the Hendrix Cut in Washington via the Great Northern. Five hundred tons were stored at the smelter. And of course, the railways took the blister copper to New York for refining and the recovery of the traces of gold and silver.

Narrow gauge railways were important at the smelter

and were integrated into the production process. An electric railway system was used in charging the furnaces, with a Baldwin-Westinghouse 30-horsepower locomotive and three charging cars serving each pair of furnaces. Typically these motors would charge the furnaces with 1.25 tons of coke and 10 tons of ore every 20–30 minutes. Slag was hauled by 14-ton Davenports, or other small steam engines, each pulling trains of six slag cars, three cars being used for each side of the furnaces. In peak production six engines and 52 slag cars were employed. The smelter had a large locomotive shed and carshop for maintenance of the railway equipment. However, in 1911 a conveyor system was installed for handling granulated slag that simplified the operation and eliminated most of the rail haulage. Simply put, the smelter was being buried in its own slag and needed another system; over the next few years the smelter became surrounded by mountains of slag.

The smelter was a dramatic sight in full operation. Acrid smoke from the smelter and its furnaces is being carried by the wind across the surrounding hills and down the Granby River to Grand Forks, devastating the fragile dryland vegetation in its wake.
–J. McRae photo, Wilson Collection, Okanagan Archives Trust.

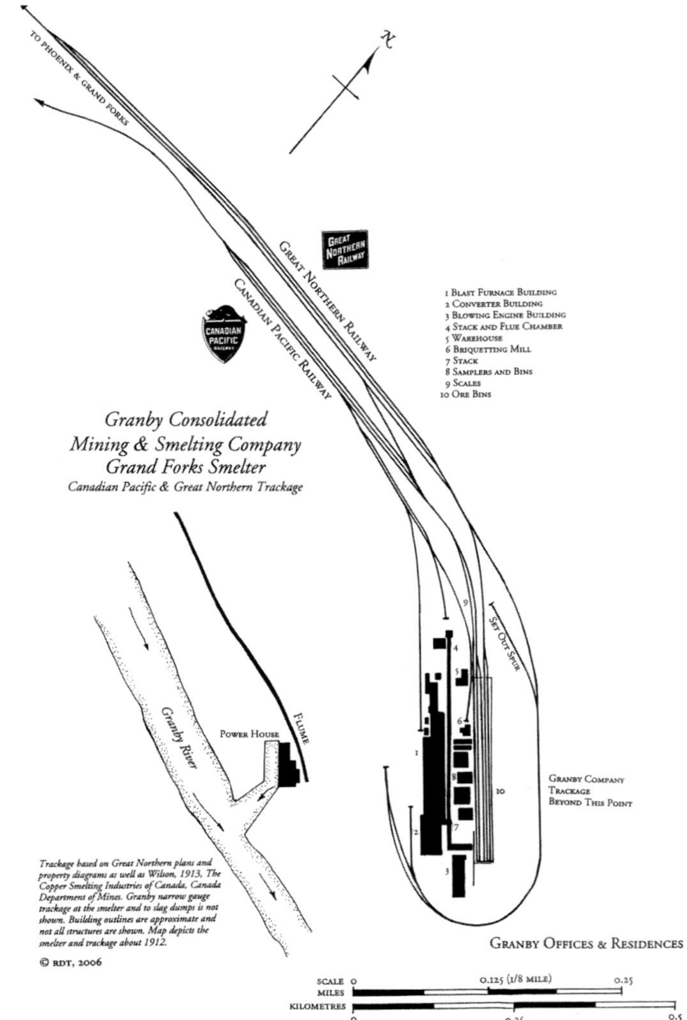

The Granby Smelter...

The huge smelter complex at Grand Forks was key to processing the copper ore from Phoenix and the silver-gold ores from Republic. Above left is an overview of the yards and the smelter, looking south with the Great Northern in the foreground and the CPR to the right. In the distance along the hillside is the right-of-way graded by the Columbia & Western but abandoned due to the demands of property owners and a dispute with Grand Forks. At left are some of the workers in the machine shop, skilled men who kept the huge plant functioning. At right is the blower room, with its massive machinery, which produced the air flow into the furnaces creating the "blast."

–THREE PHOTOS, BOUNDARY MUSEUM, NOS. 991.005.249, 998.035.001, AND W. J. CARPENTER, OPPOSITE.

PHOTOGRAPH. BY. W.J.CARPENTER. SPOKANE. WAS

Railways at the Granby Smelter...

Railways were critical to the operation of the smelters. They fed the furnaces, dumped the slag and brought in the ore from the mines. These photos show some of the railways at work at Granby in the early years.

Two small Davenports are at work dumping slag from the smelter about 1908. Each engine usually handled slag from two furnaces. Immediately above is Sherbrooke locomotive No. 2 in 1908 with two cars of molten slag at the dump or tip. The engineer was Bill Euerby. Details of Granby's locomotives are obscure but included Davenport builder's numbers 71, 72 and 75 in 1903, and 126 in 1905. All were 0-4-0 saddle-tankers.
–DAVE WILKIE COLLECTION.

Above left, a CPR crew poses beside their elderly Baldwin 2-8-0 No. 1319 at the weigh scales of the Granby Smelter. A locomotive and crew were kept busy shunting cars of ore at the weigh scales and to the ore bunkers and doing other switching work around the smelter.
–BC ARCHIVES, 44232.

Four of Granby's tank engines are at work, at left, pulling slag cars from the furnaces at Grand Forks. At right, engine crew and smelter workers pose beside Davenport locomotive No. 3 in 1908 while working between the smelter and the slag dump. Bill Euerby is the engineer in the cab of this little saddletank engine.
–W. J. CARPENTER, LEFT; DAVE WILKIE COLLECTION AT RIGHT.

Molten copper matte pours from the furnace into ingot moulds on special rail cars at the Granby Smelter. Another view, at right, shows one of the huge rotating cauldrons about to pour the copper. Below right, electric narrow gauge railways were used to charge the furnaces with the coke and ore mixture to begin the smelting process. This mine motor, with a charge car, is a Baldwin-Westinghouse product. –Above, Boundary Museum, 986.191.018; at right above, Dave Wilkie Collection; below right, BC Archives, 44668.

The Greenwood Smelter...

William Jefferson Carpenter was a master at making panorama photographic images. He captured the Greenwood smelter about 1906, and is described on the next page. This is the left half of the panorama; the right section, which shows the prosperous city of Greenwood, is reproduced on page 107.
–Greenwood Museum.

The Greenwood (Anaconda) Smelter: British Columbia Copper Company, Ltd.

The British Columbia Copper Company's smelter at Greenwood was an impressive and complex operation and was one of the cornerstones of the district's copper mining boom. It was located about one mile (1.6 km) south of the city on a terrace above Boundary Creek and directly above the CPR's Columbia & Western line to Midway. This large smelting operation was incorporated in West Virginia in 1898 and operated under licence as an "extra provincial company." The B.C. Copper Company owned extensive mining properties in the Boundary District including the Mother Lode near Deadwood, four miles (6.5 km) from Greenwood. Like many mining ventures, its origins were complex, but in 1896 a Col. J. Weir of New York formed the Boundary Mines Company and bonded several key mining claims including the Mother Lode. In 1898 the British Columbia Copper Company was established to take develop these properties. By 1912 the company owned 63 percent of the stock of the New Dominion Copper Company, and other properties. The general offices and most of the corporate officers were based in New York City.

Paul Johnson designed and then supervised the building of the smelter. Construction began in 1900 and the first furnace was blown in on February 18, 1901

Alfred W. G. Wilson described this view in 1913. "On the left... the converter building is prominent by reason of its corroded roof; behind this the down-combers of the three furnaces are visible beneath the roof of the furnace building. The buildings on the right are the power houses, while the main flue leading to the main stack on the hill is... in the background."

–Alfred W. G. Wilson, *The Copper Smelting Industries of Canada*, 1913.

(another source says February 17).[7] On average, 360 tons were smelted each day. In its first year the smelter processed 117,007 tons of ore, which yielded 3,714 tons of copper matte that contained 45–60 percent copper. In addition, the matte assayed 2–6 ounces (57–170 g) of gold and 10–300 ounces (283–8505 g) of silver per ton. Another furnace was added in 1902, giving a capacity of 700–800 tons a day. The copper ores came primarily from the Mother Lode Mine, and an analysis in 1901 showed three major types of ore. One was rich in iron (at 32.7 percent iron, with 2.8 percent copper, 5.6 percent lime and 3.7 percent sulphur), the second was richer in lime at 19.8 percent and lower in iron at 14.7 percent, along with 5.3 percent sulphur and 2.2 percent copper, while the third type of ore averaged 2.7 percent copper with 17.5 percent iron, 16 percent lime and 13.7 percent sulphur. They contained small quantities of gold and silver as well as other materials. They were described as being "exceptionally free from arsenic and antimony."[8]

Ore came largely from the company's own mines: the Mother Lode, just four miles (6.5 km) away over the CPR's Mother Lode Spur; the Rawhide, 17 miles (27 km) away; the Oro Denoro and Emma group on the line to Phoenix; the Wellington group, a few miles south of Phoenix; the Lone Star reached by five miles (8 km) of

The Greenwood Smelter Layout...

"The works are laid out in terraced plan. The power required is obtained from the Bonnington Falls power plants, through the Greenwood transformer station. The smeltery is fully equipped with ore storage bins, sample mill, blast furnace plant, converter plant, briquetting plant for flue dust, machine, blacksmith, and carpenter shops, power house, and an efficient water supply for all purposes. . ."

–Alfred W. G. Wilson, *The Copper Smelting Industries of Canada*, 1913, p. 121.

Greenwood Smelter Locomotives...

1 B-B 35-ton Westinghouse steeple cab electric

1 B+B coupled pair of 15-ton electrics

4 4 wheel electrics, 7.5 tons, 220-volt system.

Eight-car trains were used from the larger furnaces; six-car trains from the smaller furnaces. Details of builders are not known. The steeple cab is shown in photos on page 106. A photo also shows an 0-4-0 side tank engine in slag service before the electric system was installed, but details are unknown.

–The drawings of the Greenwood Smelter are reproduced from Alfred W. G. Wilson's study, cited above.

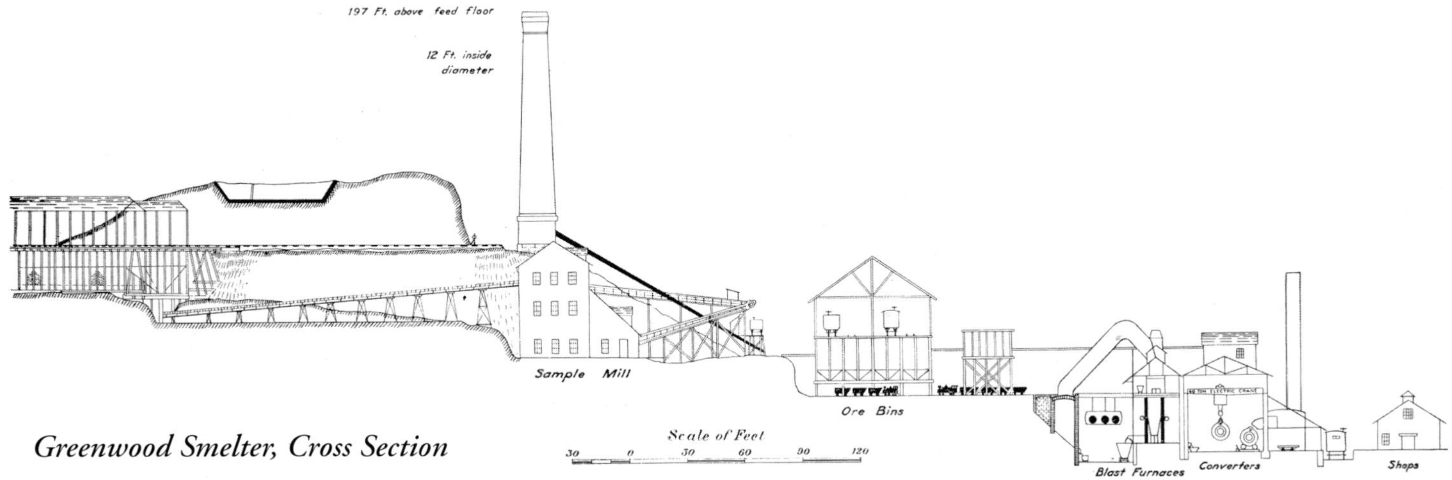

Greenwood Smelter, Cross Section

aerial tram and four miles (6.4 km) of railway; and the Napoleon group, 60 miles (over 95 km) distant on the Great Northern. Coal came from the International Coal & Coke Company in the Crowsnest, delivered by the CPR.

Electric power was supplied by the hydro plant at Bonnington Falls on the Kootenay River between Nelson and Castlegar. Auxiliary power, if needed, came from the hydro plant at Cascade on the Kettle River. Initially slag was taken to the dump using a small steam locomotive and five-ton cars. A dominant feature of the smelter after 1903 was its impressive brick smokestack towering 121 feet (36.9 m) high, most of which is still standing at Greenwood more than a century later. Situated partway up the hillside behind the smelter, it produced an effective draft of 197 feet, 6 inches (60.2 m) and was said to contain 225,000 bricks. Flue dust from the blast furnaces was partially recycled through a briquetting plant. About 1.5 percent of the materials delivered to the furnaces was recovered and converted to briquettes at a rate of about five tons an hour and then used to charge the furnaces. But the top of the huge stack was still well below the surrounding hills, and on winter days when there were temperature inversions in the valley, the potent smelter fumes could envelope Greenwood and its citizens.

In 1904 a Bessemerizing plant was built, enabling the smelter to produce blister copper instead of the much less pure matte, substantially reducing the volume of shipments and costs to the refinery. In 1906 the two original furnaces were removed and three new and more efficient ones were installed. The plant was enlarged again in 1907 to give a capacity of 1,700 tons a day and again in 1910 by another 500 tons a day. The slag removal system also was upgraded with the use of an electric railway and 25-ton steel slag cars that moved the molten slag to an expansive dump to the southwest of the smelter.[9] A century later, this huge dump and the smelter stack remain as two of the most visible reminders of the copper smelting industry that once dominated the landscape and economy of this region.

Ore and other materials were delivered to the smelter complex by the CPR to three levels of the massive plant. The upper level was for ores needing sampling; the middle level received coke and previously sampled ores; while the bottom level was organized to receive supplies, to remove slag from the smelter and to ship the processed blister copper. Ore bins at the upper level held 2,000 tons, the intermediate-level ore bins held 12,000 tons and the coke bins could store 3,000 tons.

Paul Johnson designed and oversaw the construction and operation of the Greenwood smelter. The smelting works are shown above and in plan view below. In the foreground of the photo is the CPR's covered bridge over Boundary Creek.

–*The Mining Record*; Mines Branch, Department of mines at right.

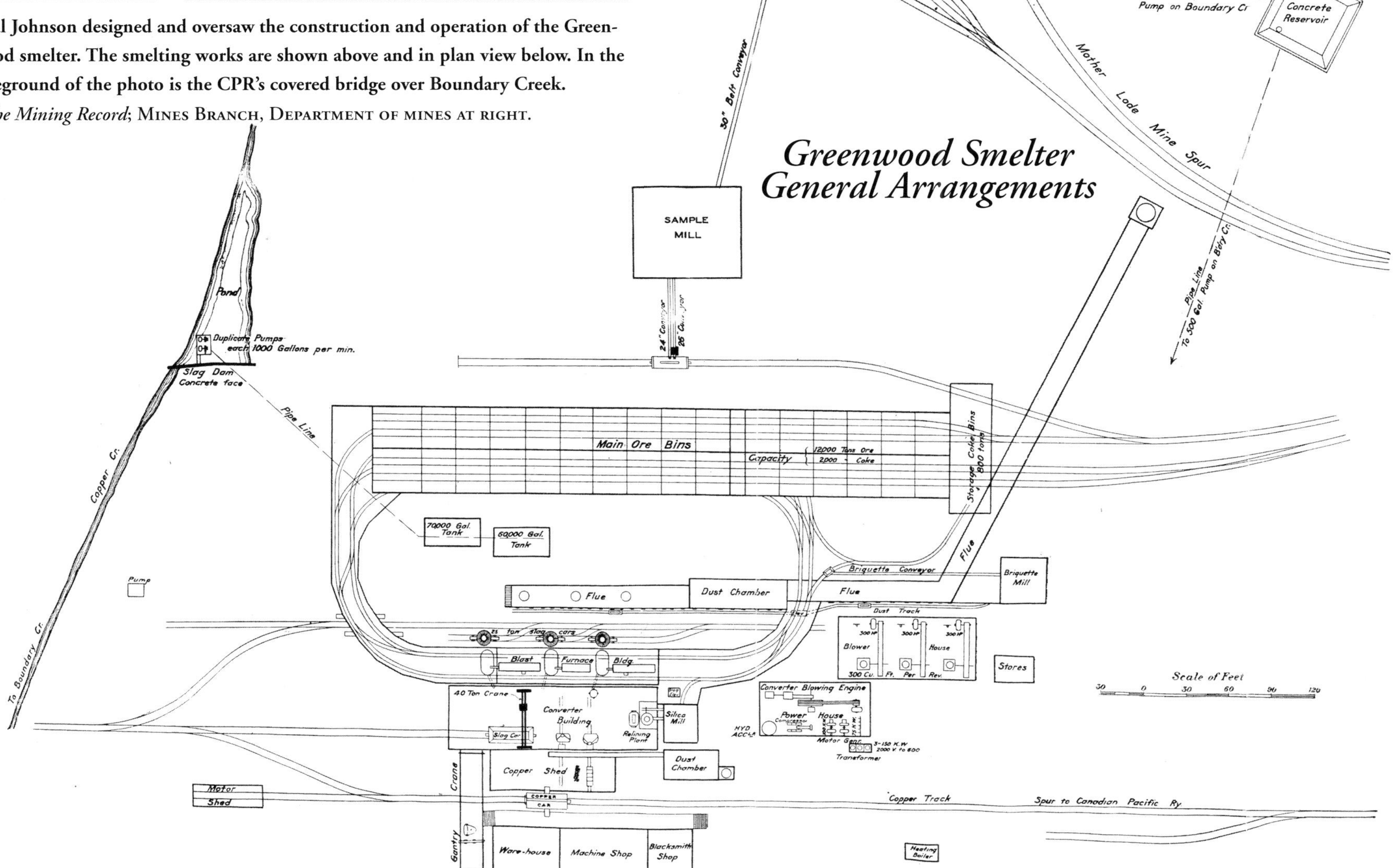

Dumping molten slag at Greenwood using the electric railway system, and leaving a landmark that would remain for over 100 years.
–Greenwood Museum.

Greenwood Smelter Labour Force...

"Open shop is the policy of the smelter. To operate the plant at its present capacity, the following force is required per 24 hours: superintendents and warehouse clerks, 4; shift bosses, 5; sampling mill, 11; blast furnaces and briquette mill, 61; converting department, 15; powerhouse, 3; mechanical and electrical force, 12; slag dump and general labour, 9; total 120. The plant is thus treating approximately 20 tons of ore per day to each employee, including superintendents and foremen. The sample mill staff, and the men engaged around the furnaces and converters work 8-hour shifts; the major portion of the other labour works 9-hour shifts."
–Alfred W. G. Wilson, *The Copper Smelting Industries of Canada*, 1913, p. 132.

The inside of the massive smelter was noisy, dirty, hot and dangerous. Above, the interior of the blast furnace building on what was called the matte side, and at right, the slag side. No hard hats, no safety goggles, no safety gloves, no workers' compensation in case of injury; only the support of other workers, a lodge or the union.
–Mines Branch, above; Greenwood Museum, right.

Within the plant an extensive railway system was used to move ore, slag and coke. In 1913, three 7.5-ton electric locomotives, operating on a 220-volt system, were used for hauling the charging trains and a fourth locomotive was held in reserve. Twenty-six side-dump cars with a capacity of 55 cubic feet (1.5 m^3) were used for charging the furnaces. The charging trains, operated by a motorman and two loaders, were usually six or eight cars in length and one was operated to each furnace. They ran to any of the ore bins to collect an appropriate mix of ores, coke or briquettes and then moved to the blast furnaces to unload the charging materials. Another electric railway system was used to dump the slag. For this, one 35-ton motor and a coupled pair of 15-ton machines were available.

The cupola building of the smelter, which contained the three blast furnaces, was 60 x 150 feet (18 x 46 m). The adjoining converter building was 44 x 150 feet (14 x 46 m) and the powerhouse was 40 x 81 feet (12 x 25 m). All were of steel construction. The blast furnaces originally were each 48 x 240 inches (1.2 x 6.1 m) at the tuyères, but the end furnaces were enlarged to 51 x 360 inches (1.3 x 9.1m), increasing efficiency.

Copper matte was poured from the furnaces into five-ton ladles. It was then moved to the converters for further processing. There were two converter stands and five extra converter shells in the converter facility, and two charges of blister copper could be made before the lining of the converter had to be replaced. These linings used siliceous gold ores brought from either Republic or the Snowstorm Mine in Idaho. This plant could handle 40 to 50 tons of 35 percent matte every 24 hours or about 1,000,000 pounds (453 592 kg) of blister copper a month. Slag from the converter was poured into sand beds to cool, then moved by rail car to the ore bunkers where it was stored for recharging the blast furnaces so that any additional copper could be recovered. Finally, the blister copper was cast into slabs, weighed and shipped out by railway to the eastern states for refining into pure copper suitable for industrial and commercial uses, in a world that was demanding electricity, telephones and many other services.

The city of Greenwood was situated in the valley of Boundary Creek just upstream from the smelter. W. J. Carpenter's panorama captured the smelter, as shown on pages 102 and 103, and continued north and east to frame the bustling community.
–W. J. Carpenter photo, Greenwood Museum.

The locomotives used on the Boundary mining branch lines were rugged, often unforgiving machines. CPR's 953, a Baldwin 2-8-0 built in 1896 and shown at Greenwood in 1903, was typical with 49-inch drivers and 20x26-inch cylinders. These were "deckless" engines, and the crews sat beside the boilers, where a broken steam pipe could be deadly. Engineer John W. Goostrey is third from left and Martin Anderson is standing at his left.
–Greenwood Museum.

The Boundary Falls Smelter...

The Boundary Falls smelter had at least two small locomotives used in slag service on a 42-inch (1067-mm) narrow gauge system. Both were 0-4-0Ts built by Davenport in 1906 as shop numbers 166 (3-1906) and 509 (9-1906). The detailed views at right show these two engines at work. Their fate and other details are unknown. The large smelter never became a profitable concern and only operated intermittently. A century later little remains at the site except for remains of the slag dumps, but they are minor compared to the huge accumulations at Greenwood or Grand Forks.

The Boundary Falls Smelter:
The Standard Pyretic Smelting Company
Montreal & Boston Copper Company
Dominion Copper Company
New Dominion Copper Company

This was the least successful and productive of the three smelters associated with the boom at Phoenix and the surrounding mining areas. The origins of this operation are twofold. One dates to the 1890s when Mackenzie and Mann acquired substantial interests in several of the key mining properties in the district and the next year the Dominion Copper Company was formed.

Construction of the Boundary Falls smelter itself began in 1900 and was completed the following year by the Standard Pyretic Smelting Company. It was soon purchased by the Montreal & Boston Copper Company, which operated it until May 20, 1905. Later that year it was acquired by the Dominion Copper Company, which ran it intermittently until 1908 when the company went into receivership, then in 1909 the properties were sold. They were reorganized under the name of the New Dominion Copper Company (in which the British Columbia Copper Company had a controlling interest), and they refurbished the plant, including reconstructing the smelter's power plant that had burned down. But despite these investments the smelter was not reopened and ores from the mines controlled by the company, which included the Brooklyn, Idaho, Standard, Stemwinder, Montezuma and Rawhide mines, were sent to the nearby B.C. Copper Company smelter at Greenwood which was better able to smelt the ore efficiently.

The smelter, sometimes called the "Sunset Smelter," originally had three water-jacketed blast furnaces with a capacity of about 300 tons a day. In 1907 a blast furnace of 600-ton capacity was installed, but it saw only limited use before the plant closed in what proved to be the end of smelting at this once impressive facility and symbol of mining prosperity.[10]

Meticulous in his work, photographer William Carpenter set up his panoramic camera to record this scene of the then busy smelter at Boundary Falls. The two detail views on the facing page, showing a locomotive taking a car of slag to the dump and another engine with a badly dented saddletank, are enlarged from this panorama. These early panorama views produced a characteristic curving distortion of the image that was a result of the extremely wide field of view taken in by the special camera and its lens.
–W. J. Carpenter photo, Greenwood Museum.

Shay Tests...

"The tests of the Shay engine have not been completed, and it has not been in service a sufficient length of time to warrant our constructing a line to be operated exclusively by this class of power; but we are fairly well satisfied that it will be a success for heavy grades."

–Thomas Shaughnessy to W. F. Tye, October 15, 1900.

At Trail and Phoenix...

"The big Shay engine that plied for many months between Smelter Jct. [Trail] and Rossland has been sent to Eholt says the *Rossland Miner.* Or rather will be taken there today by Conductor Joe Kenny. She will be utilized to haul ore on the hill out of Phoenix a run of about ten miles similar in almost every respect to the Rossland Hill. As the Shay was operated most successfully here it is only natural to suppose that the big mill will be equally successful at Phoenix. When mines on Red Mountain recommence shipping ore No. 111 will be brought back or the run will be taken over by a new Shay recently ordered by the CPR."

–*The Phoenix Pioneer*, August 31, 1901.

The CPR's Shays, Their Most Powerful Locomotives

The heavy mining traffic and steep grades to Phoenix as well as those between Rossland and Trail presented Canadian Pacific with challenges because their existing conventional locomotives could handle only modest tonnages. At that time alternatives generally were limited to some early articulated engines like Fairlies (and no evidence has come to light to suggest that the CPR considered using them), very expensive and still somewhat experimental electrification of the mining sections of the railway, or the use of slow but powerful geared locomotives. Interestingly, Canadian Pacific did look into electrification schemes for parts of its main line and mining branches in the years just before the First World War. But at the turn of the 20th century, the Canadian Pacific chose to try out geared locomotives in the form of Shays, and for a time they were the most powerful engines on the CPR system. The first, No. 111, was delivered from the Lima Locomotive & Machine Company in Lima, Ohio, in April 1900 and was unusual enough to warrant a brief article in *Scientific American.* Two more followed: one in May 1902 and the last in December 1903.[11]

The first Shay had 15 x 17-inch cylinders and 40-inch drivers. In working order it weighed 124 tons, and as the late Omer Lavallée, the CPR's distinguished historian and locomotive authority, pointed out, was "rated at 225 %, [and was] by far the most powerful locomotive on the system at that time." As well as being a very powerful locomotive, the short wheelbase of the Shay's trucks gave it the ability to operate over rough track and around tight curves. Only three were ordered for the CPR mining branch lines. The first two Shays were numbered 111 and 112, but before the arrival of the third, CPR renumbered its locomotives and the Shays became 1901, 1902 and 1903. In the system-wide renumbering of 1912, the engines became 5901 and 5903, the second Shay by then having been scrapped in April 1911.

When first tried out between Trail and Rossland, Shay 111, it was claimed, could haul 312 tons against 130 tons from other locomotives then in use at Trail. Soon the Shay was moved to Phoenix, but the delivery trip was not without incident. Near Porcupine Creek bridge on Farron Hill, the locomotive struck a rock and broke a connecting rod, but the crew was able to coax the engine to Grand Forks where an engine from Eholt met the train. After a long struggle, and "endless delays," they reached Phoenix with the Shay, 23 ore cars and 50 passengers (presumably riding in a coach). It was an unfortunate minor incident, but at least the engine was easily fixed.[12]

Unfortunately, information on the operation of the Shays is limited. In June 1902 two Shays were in service between Phoenix and the Granby Smelter, reportedly handling 25 cars downgrade and 20 empties back to the mines, said to be twice the capacity of a standard engine. However, a report noted that Shay 111 was being returned to Trail near the end of August 1902 because of its slow speed on the grades. Yet, CPR assignment records show all three assigned to "Granby, B.C." on July 31, 1904, and working in freight service either at Phoenix, at Granby or on the Mother Lode branch. The 1901 was at Phoenix in August 1904 when it was wrecked working a regular ore train of 24 loaded cars, one empty car and a caboose at Mile 2.75 on the branch line. The pileup was spectacular, and fortunately the crew, including engineer Al Kenward and fireman Charles Haggitt, jumped in time, but *The Grand Forks News-Gazette* claimed the CPR's loss was $150,000. However, the engine was salvaged and rebuilt at Nelson, and the ore was recovered. The paper reported that "over 700 pictures of the wreck have been taken [sold]," by the photographer.[13] A 1910 report noted all three Shays being used in ballasting service on the Phoenix Subdivision which was being rebuilt with heavier rail and improved ballast. This may have been the way they were used in their later years of service. Shay 111 (1901) also operated on the Mother Lode branch from Greenwood to the nearby mines.

The Shays were replaced by heavier standard locomotives, usually 2-8-0s, that were simpler to maintain and operate. In the early 1900s, the CPR ordered many of these engines, larger and with greater power, and they were widespread throughout the CPR system. The Shays' strength on steep grades, derived from their complex gearing, was also their weakness because it mandated slow speed. A Shay would normally not operate above 10 miles an hour (15 km an hour) even on level tracks, where a standard locomotive regained much of its power and could operate at substantially better speeds. Maintenance records for 1904, not counting the period after 1901's wreck, show the engines out of service 20 percent of the time, which was about twice as long as 2-8-0s in the same service, and may well have indicated the underlying problems with them. It is not known when the Shays last operated at Phoenix or Rossland, but they all had short lives on the CPR. Shay 1901 was scrapped in February 1914, 1902 having preceded it in April 1911, while 1903 was sold in August 1913 to the Corbin Coal & Coke Company in the Crowsnest Pass for work around the open-pit coal mine just west of the summit of the pass.

The CPR's Shays were impressive locomotives, and sported classy wooden pilots. Opposite, the 111 when new and ready to enter service. In panorama, it is in ore service at Phoenix, and immediately above, one of the CPR Shays is at Eholt in 1910.
–Canadian Pacific Archives, at left; Cominco Collection top; and BC Archives, 51802 (detail).

The Great Northern Expands and Builds to Phoenix

James J. Hill, the Empire Builder...

"His iron grey hair was a little long, and covered with a plain brown cap such as an ordinary impecunious mortal might wear. A dark grey suit and blue and white striped shirt with a plain necktie complete the outfit. A stranger would hardly suspect in this thick-set, quiet-voiced and very plainly dressed man the personality of one of the most brilliant railway men of the continent.

"The Gazette representative recalled to his memory his first visit into the Boundary. 'There have been things doing since that time, Mr. Hill.' 'Well, yes. I like very much better the doing of things than the promising,' and the reply was accompanied by one of those sudden, humorous smiles which subtly alter the whole look of that dark and sometimes grim face.'"

—*The Grand Forks Gazette*, SEPTEMBER 30, 1905.

Building to Phoenix...

"Over 1,000 men were reported to be at work on the construction of the line between Grand Forks and Phoenix, B.C., including a two-mile spur to the Granby mines early in July, and P. Welch stated that the number would be doubled by the end of the month."

–*The Railway and Shipping World*, AUGUST 1904, P. 279.

IN THE EARLY 1900S, THE GREAT NORTHERN WAS advancing on several fronts in its drive to increase revenue from the mining districts adjacent to its main line. Hill saw clear advantages in coordinating the production and transportation components of the mining industry, as did of course his rivals in the CPR. Not only had he acquired interests in the Grand Forks smelter, but he had also purchased a major interest in the Crow's Nest Pass Coal Company and built a branch line from Jennings, on the GN main line through northern Montana, to Morrissey and Fernie in the heart of the Crowsnest coalfield in British Columbia. Crucial to the smelting processes and the operation of the mining districts was a reliable supply of coke and coal. Moreover, coal was a vital commodity for the operation of the railroads because nearly all locomotives at that time depended on coal for fuel. This was a long-term need by the Great Northern and key to several of its most important shippers. Moving coal for locomotive fuel did not produce revenue directly but a good supply of cheap coal could save substantial sums and was certainly preferable to paying another company such as the CPR to haul the coal over its lines to an interchange with the Great Northern. The linkage between coal and coke, and the mining and smelting industries, was intimate and essential.

Hill's interest in the Crowsnest Pass coalfield went back to the late 1880s and the construction of the GN main line in the early 1890s. In 1891 he had investigations made of the area and later wrote to Van Horne that he thought that coal from the Crow could be used until Washington deposits were developed.[1]

Hill secured the charter of the Crow's Nest Southern Railway that enabled the Great Northern to build a line from the Crowsnest Pass in a southwest direction to the B.C.–Montana border, there to connect on south to the Great Northern's main line at Jennings using the charter of the Montana & Great Northern.[2] The Great Northern tracks were opened to Morrissey, the farthest east in British Columbia of the most important mining and coke-producing centres, in 1902, within four years

A. B. W. Hodges replaced William Yolan Williams as superintendent of the Granby Smelter in 1904 when ownership shifted to New York based interests. Hodges had overseen the construction of the smelter itself and its initial operation and expansion. He had training as a metallurgist in the United States.
–*The Mining Record*, 1901.

Photo, facing page...
Great Northern's Phoenix branch featured this impressive wooden trestle, Bridge 66, over Deadman's Creek (later called Glenside Creek) at Mile 18.2. It was 195 feet (59.5 m) high and 672 feet (205 m) long and built on a 14-degree curve. This is probably a construction train at the time of the line's opening between Grand Forks and Phoenix. Other views of this trestle are on pages 118 and 120.

–LIBBY PHOTO, EASTERN WASHINGTON STATE HISTORICAL SOCIETY, L94-40-60.

Great Northern operations to Fernie, Michel and the Crowsnest mining centres were impressive. In busy times 100-car trains, sometimes powered by big Mallets like L-2 Class 1831, shown at Fernie in 1911, moved coal and coke south to the GN main line at Rexford, Montana. This locomotive, built by Baldwin in 1908. The photo at right shows the coal processing facilities at Michel in the Crowsnest Pass.

–National Archives of Canada, c-21008, above; Dave Wilkie Collection, at right.

Coal from the Crow...

"The CPR if it is to successfully compete with the Great Northern for coal tonnage from this district will have to greatly improve the grades on the Crow's Nest line. Fourteen cars is a train on the CPR while the GN easily takes out 60 cars of coal of 20 tons each."

–*The Grand Forks News-Gazette*, February 21, 1903.

of the opening of the Canadian Pacific's Crowsnest Pass route. Two years later, the tracks were extended 9.8 miles (15.8 km) to Fernie and a further 17.7 miles (28.5 km) to Michel in 1908. Typical of Hill enterprises, the new railway was built on an excellent route with easy grades to the mines. The line followed the Kootenai/Kootenay River northward and then ascended the Crow by following the course of the Elk River.[3]

In 1903–1904, to reduce grades on the GN main line through northern Montana between Jennings and Columbia Falls, a new line was built between the original main line at Columbia Falls and Rexford. There it connected with the southern end of the new branch line to the Crowsnest, forming a new route through the mountains and, at the same time, bringing the Crow and its essential coal and coke that much closer to the GN main line.

For the Great Northern, needing coal for its locomotives and for the industries developing along its railway, Crowsnest coal and coke became very important, and in 1906 Hill and his associates expanded their stock holdings to a controlling interest in the Crow's Nest Pass Coal Company.[4] In 1907, Hill wrote to George A. Cox, president of the Crow's Nest Pass Coal Company, that "our demand at present... is 2,500 tons daily [as much as 80–100 carloads]—more when our new line from Spokane to Portland is completed."[5] In addition were the needs of the mines and smelters along the GN in Montana, Washington and southern British Columbia. At times, the demand for steam coal and coke exceeded production and shortages of coke were a serious concern for the smelters in the Boundary District, curtailing production. In mid-1907, for example, coke was in such short supply that the Boundary smelters had to shut down some blast furnaces. "Much complaint is heard of the Crow's Nest Pass Coal Co. shipping large quantities of coke to smelters on the American side, contrary to the terms of its charter, when the coke is really required in British Columbia," reported *The Canadian Mining Journal*.[6]

The comparatively gentle grades on the GN's Crowsnest Pass lines gave it a considerable advantage over the CPR. Once again, Hill's determination to build railroads with low grades was demonstrating its wisdom and value. As coal production increased and as larger locomotives were assigned to the GN branch, 100-car trains could be hauled south to Rexford.[7]

The construction of trackage into the Crowsnest coalfield provided yet another area of expansion of Great Northern interests in southern British Columbia. At the same time, Hill was sabre-rattling about other expansion in Canada. At this time the Great Northern was in excellent financial shape and was moving to consolidate its various railroad interests. Following a

Minnesota Supreme Court decision in 1898 permitting better financial returns on its investments, substantial new capital investments followed new share offerings by the Great Northern and over the next 10 years the GN consolidated its finances and the operations of its various properties including the VV&E, the Midland Railway of Manitoba, the Brandon & Saskatoon and other enterprises. All this added fuel to Hill's enterprises and increased speculation, which he happily encouraged, about his intentions north of the border. A news report noted that $10,000,000 from new stock offerings would be applied to expansion in western Canada. "The system to be projected will be the equal of the present Hill lines in the United States. Winnipeg will be the Canadian headquarters and vast terminals are being secured or have been purchased in Winnipeg, Minneapolis, Chicago, Vancouver and Seattle. Winnipeg, Brandon, Regina, Edmonton, Prince Albert and Peace River valley will be ultimately covered by the Hill system." Another report had the Great Northern moving into the vast wheat lands of Saskatchewan under the name of the Saskatchewan Central Railway and noted that a bill authorizing construction was being considered by the Saskatchewan legislature. According to the hyperbole, a high-class freight railway would be built through to the Pacific and, of course, elements of the route were taking shape but not with the vigour the reports suggested. [8]

Meanwhile, the CPR was not idle. It struck an alliance with Edward Henry (E. H.) Harriman's Union Pacific, with which Hill was engaged in competitive struggles along the Columbia. The CPR invested in the Spokane International Railway, and secured an option to buy a controlling interest in the line. This railway was built by Daniel Chase Corbin between the CPR Crowsnest line at Kingsgate, on the B.C.–Idaho border near Yahk, B.C., and Spokane with connections over the Union Pacific to Portland, giving Canadian Pacific a new route to compete with Hill's main line.

The SI, as it was often called, provided a new international gateway. [9] Moreover, the CP/UP route gave the CPR another way to compete directly with the GN for traffic from the Midwest and the Pacific coast, and to underline the fact, a new train, the *Soo-Pacific Train Deluxe,* was placed in service in 1906.[10] However, Hill, publicly at least, was not impressed, as *The Wall Street Journal* reported in July 1907. Rhetoric high, he told the CPR to "run all the passenger trains [it wished] between St. Paul and Puget Sound. Put on conservatories, flowers, and music. If there is any big deficit I might help to make it up; but keep out of my freight preserves or there will be trouble. There's traffic up in the Canadian plains waiting for me and I am going to have it!"[11]

The Great Northern interests were being developed with careful political and strategic regard for the international boundary, demonstrating Hill's conviction that free trade between the two countries was essential, inevitable and would be beneficial to business. In contrast, in engineering terms, the expanding GN lines effectively ignored the boundary as the survey engineers worked out the most effective routes for the branch lines between the mining districts, and crossed and recrossed the international boundary as needed.

While the Great Northern was building into the Crowsnest Pass, a territory that the CPR had once considered its own, large-scale developments were progressing in the Boundary District. On June 20, 1904, the first car of blister copper was shipped from the B.C. Copper Company smelter at Greenwood. Yet another smelter, this one owned by the Dominion Copper Company at Boundary Falls, just a few miles south of Greenwood, came into production that fall, and on September 18, 1904, the first trainload of ore was delivered there.[12] This brought to three the number of smelters in the Boundary District in addition to the nearby smelters at Trail and Northport on the Columbia River.

Construction work began in earnest on the Great Northern's Phoenix branch in mid-1904 when about 500 labourers started clearing the right-of-way; another 1400 were employed on grading in June. John W. Stewart was the general contractor and Patrick Welch the manager for the 28 miles (45 km) of main line and sidings required to reach the mining district. The Porter Bros. won the

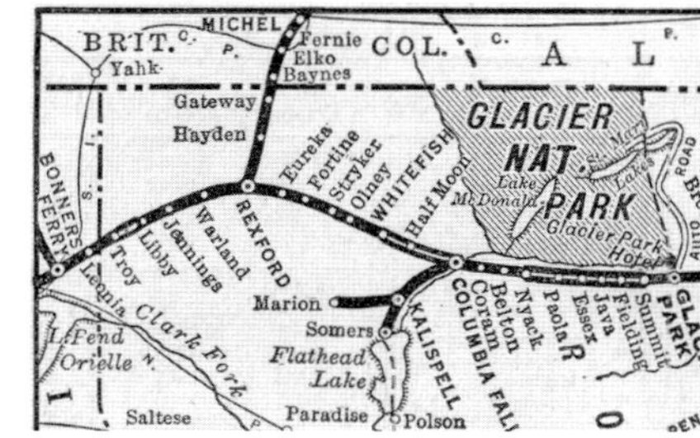

228 Mixed Ex. Su.	Mls	Table 69 Mountain Time		227 Mixed Ex. Su.
5 40	0	Lv **Rexford** *6, 11*	Ar	**1 00**
f 5 50	5	" Hayden	Lv	f**12 45**
6 00	9	" Gateway, Mont.	"	**12 35**
6 10	9	" Internat'l Bdry.	"	**12 23**
f— —	9	" Newgate, B. C.	"	f— —
f 6 25	16	" Flagstone, B. C.	"	f**12 01**
f 6 37	19	" Dorr	"	f**11 49**
6 47	24	" Waldo	"	11 34
f— —	25	" Baker Spur	"	f— —
7 05	30	" Baynes	"	11 16
f 7 22	37	" Mott	"	f10 54
7 33	42	" Elko	"	10 40
8 00	52	" Swinton	"	f10 15
8 20	62	Ar **Fernie**	Lv	9 50
....	62	Lv **Fernie**	Ar	....
....	69	" Hosmer	Lv	....
....	73	" Olson	"	....
....	82	" **Michel**	Lv	....

REX ORD, FERNIE AND MICHEL

Great Northern passenger service from Rexford, on the GN main line, to the Crowsnest Pass coal mining centre of Fernie was down to a daily except Sunday mixed train by October 1, 1924, when this timetable was effective. To cut expenses on these marginal services, Great Northern introduced gas-electric motorcars, sometimes called "gas-cars," like those shown on pages 176 and 177, to many branch lines including the one to Fernie. Freight trains were run as frequently as conditions demanded, and the low level of passenger service does not suggest an unproductive line.

–Robert Turner Collection.

At Phoenix the Great Northern built a wye using two large wooden trestles and a truss bridge. This scene, most likely of a construction train, shows Phoenix probably in January or February of 1905 just before the railway started hauling ore to Grand Forks.
–Libby photo, Eastern Washington State Historical Society, l94-50.61 (restored).

Above right, former Spokane Falls & Northern No. 2 at Phoenix in 1905 on the new spur to the Granby No. 3 tunnel ore bins. This Baldwin 2-6-0 was built in 1889 and became GN's No. 479 in 1908. It worked on the Phoenix passenger service between 1905 and 1918, although sometimes ore was handled from the Idaho Mine between Phoenix to Denoro. and then 700-class GN 4-8-0s were used.
–Dave Wilkie Collection.

contract for all the bridges on the Phoenix line and for the spur to the Granby Smelter that required a 1000-foot (305-m) trestle over the CPR. Subcontractors soon had seven construction camps set up along the right-of-way to Phoenix. Further contracts were let in October for the switching yards and locomotive facilities including a roundhouse, coal dock and repair shops.[13]

The Great Northern route headed westward from Grand Forks and climbed steadily along the face of the mountains framing the north side of Grand Forks overlooking a vista of beautiful farmlands and orchards. The shelf for the right-of-way and the fills and slopes of rock debris from cuts along the hillsides were to endure for decades. Then the line turned north up the east side of July Creek to climb up the valley to B.C. Junction where it turned west under the CPR and, by a series of loops, entered Phoenix from the northwest. The GN depot was located not far from the Phoenix city hall, on the opposite side of town from the CPR station. Switchbacks connected the small GN yard to several of the major properties including the Idaho, Old Ironsides and Victoria mines. Some impressive trestles were required, including one 700 feet (213 m) long and 165 feet (50 m) high on a 14-degree curve spanning Deadman's Gulch (later given the more respectable name of Glenside Creek). Another long trestle was required to cross Dominion Avenue in Phoenix, and on the Coryell Ranch, where the tracks ran below rock bluffs, a 572-foot (174.3-m) trestle was needed.

Track-laying was within eight miles (about 12 km) of Phoenix by early December 1904. Anticipating more ore shipments with the GNR's pending arrival at Phoenix and the market continuing to flourish, the Granby Smelter doubled its capacity that winter. Three new furnaces were in operation by February, coinciding with the completion of the Great Northern's line to Phoenix.[14]

The total construction costs recorded by the Great Northern for this modest branch line of 27.85 miles (44.82 km) were substantial at $1,339,014.96, and averaged $48,074 per mile ($29,872 per km) by the time service began. Acquisition of the right-of-way came to $88,940. Of the overall total $749,165 was for grading, and $100,639 was for bridges, culverts and masonry, while rail, switches, ties and other track hardware cost $192,489. Track construction, grading and surfacing amounted to $73,141. Additional costs for ballasting added an estimated $750 a mile (about $460 a km).[15]

Great Northern trains began hauling Phoenix ore

With the extension of regular services to Phoenix by the Great Northern residents of the mining town had two railways from which to choose. This is the GN train beside the new depot at the western end of Phoenix. The railway was operated as part of the Spokane Falls & Northern system that the GN owned, and it later became the Spokane Falls & Northern Division. In 1909 these branch lines came under the Marcus Division, and in 1921 under the Spokane Division of the GNR. Several of the original SF&N engines were used on the Phoenix trains, particularly on the passenger service.
–Dave Wilkie Collection.

on February 15, 1905, when five cars were delivered to the Granby Smelter although ballasting of the branch line still remained to be completed.[16] Six locomotives were brought in to move the ballast trains and finish the work as quickly as possible. Unfortunately, the day before trains began running, a fire swept through downtown Phoenix, destroying the Brooklyn and Phoenix hotels, two of the city's largest, but such events were common in the frontier mining towns. Rebuilding usually followed rapidly if times were prosperous, and there was no better omen of a town's future than the extension of a railway into the heart of the business district.

Initially, the Great Northern was expected to move 10 cars or 500 tons a day but reports the next month showed that the new railway was moving as many as 28 cars from the mines each day. Grand Forks' *Gazette* reported on March 25 that Canadian Pacific had laid off three crews between Nelson and Phoenix as a result of Great Northern competition although 50 to 60 carloads a day were still being handled by the railway. Passenger traffic to the coast also was down because the Great Northern route via Spokane was a day quicker than the CPR's service via the Arrow Lakes steamer route to Revelstoke and the main line from there to Vancouver.[17] Great Northern passenger service began on March 5 with a once-a-day train to Grand Forks and return.[18] The new line meant that travellers from Phoenix could be in Spokane the same day, was cutting travel time substantially over the CPR route to the east. Many Phoenix residents had come from Spokane and there were also many business connections to the Inland Empire. Overall, tonnages to the mines and smelters continued to grow and both railways were kept busy. Nonetheless, the CPR's near monopoly, after five years, was clearly at an end.

The changes that were taking place in the Boundary District and the Great Northern's growing interests in the region prompted visits from many leading American

Unreliable Schedules...

"The change in the CPR time schedule left a considerable number behind on Monday. The train is now departing at 11:30 a.m. instead of 2:00 p.m."

–*The Phoenix Pioneer*, June 11, 1910.

On Time Was Surprising...

"The GN passenger train actually arrived in Phoenix on time Thursday evening. It was the biggest shock the local terminal had received for many a moon."

–*The Phoenix Pioneer*, November 19, 1910.

Many Bridges to Phoenix and the Smelter...

The Great Northern line to Phoenix from Grand Forks was a spectacular piece of mountain railway, with 92 bridges including several large trestles. In the photo above left, a northbound empty train of steel ore cars, hauled by a G-3 class 4-8-0 with another engine pushing, is thundering back upgrade over Bridge 66 to Phoenix from the Granby Smelter at Grand Forks. Unlike many trestles, this one had a ballasted deck. The same bridge is shown above in winter weather, with very wet-looking snow, as the train crew stopped their engine, an unidentified 4-6-0, and a four-wheel GN caboose or "bobber," near the end of the huge timber structure for the photograph. At left, the Great Northern's yard in Phoenix had a small two-stall engine house and a 50,000-gallon water tank not far from the station. The wye was built with two of its tracks on large curved trestles. The track in the foreground leads to Grand Forks, the one at right to the Granby mines and the one at left to the Great Northern depot, shown on the previous page, a short distance away, and not far from the town centre. Just past the depot was the Idaho Mine.

–BC Archives, 44666, above left; Libby photo, Eastern Washington State Historical Society, L94-40.61, above; Wilson Collection, Okanagan Archives Trust, at left.

Great Northern's line to the Granby Smelter required another impressive bridge (far left), this time over the North Fork of the Kettle River. Opened in 1905, it was built under contract by the Porter Bros., who built the bridges and trestles on the Phoenix branch and the smelter spur. Its 658 feet (200 m) included a combination of trestlework, deck truss spans and a through span that crossed over the CPR line. The bridge was just upstream of the dam and flume used as part of the hydroelectric generating system for the smelter. The dam itself required 400,000 feet of lumber. The centre photo shows the CPR's bridge after the truss span was sheeted for weather protection. The photo above depicts the CPR span in the foreground with the higher Great Northern structure behind. This view is looking west, downstream of the power plant. Both bridges were replaced by new structures during the First World War.

–Andrew McCulloch Collection, two photos, courtesy Barrie Sanford, and BC Archives, 44240, above.

businessmen. On September 27, 1905, James J. Hill arrived in Phoenix and Grand Forks from Fernie on his way west to Portland, Oregon, and Vancouver, B.C. With an entourage of some of the most influential railroadmen and bankers in America, Hill was making a grand tour of his empire, including the Northern Pacific and Burlington lines over which he and his associates, in particular J. Pierpont Morgan & Company and the Deutsche Bank, had recently gained control, and at the same time bolstering support from his financiers for further expansion.[19] Hill had few worries about being greeted by a hostile press. "The man who is generally acknowledged to be head and shoulders above any other in the American railway world—James J. Hill—made his long deferred initial visit to Phoenix...," effervesced *The Phoenix Pioneer* on September 30 in unrestrained enthusiasm for the Great Northern's 67-year-old patriarch and empire builder.

A tour of the Granby's Phoenix mine facilities and the underground and surface workings impressed the businessmen. Granby General Manager Jay P. Graves, who accompanied the party for this portion of the tour led by A. B. W. Hodges, general superintendent, and O. B. Smith, Jr., mine superintendent, took the guests through the extensive mine workings. They were shown one of the ore quarries—a glory hole—being worked by steam shovel and other new developments before taking the train down to Grand Forks.

"Out into the rain stepped the biggest aggregation of millionaires that ever sat in a railway car in the west," gushed *The Grand Forks Gazette*. "In monied worth," Grand Forks' *Evening Sun* noted effusively, "the members outclassed any other combination of men that have ever occupied the same Pullman car at one time. The party is reputed to have represented $250,000,000...." Arriving in a special plush seven-car train called the "President's Special," Hill's party included his son Louis W. Hill, who had been through Phoenix and Grand Forks just two months earlier, and many long-time Hill associates and financial backers. These included R. C. Morgan, superintendent of the Spokane Falls & Northern; Howard Elliott, president of the Northern Pacific; Jule M. Hannaford, second vice-president of the Northern Pacific; George B. Harris, president of the Chicago, Burlington & Quincy; Darius Miller, first vice-president of the CB&Q; William B. Deane, director of the Great Northern; and Benjamin Campbell, fourth vice-president of the GN. The banking contingent included Grant B. Schley of the powerful and influential New York banking and brokerage firm of Moore & Schley, who were long-time backers of Hill's enterprises; Philadelphia

Bill Euerby, Great Northern Engineer...

"The GN had ten locos at Grand Forks and ran three ore trains from Phoenix to the smelter at Grand Forks. They ran two double headers with 25 50-ton dumps, and one single [engine train]. The doubles required pushers upgrade to Phoenix. That gave a total of about 64 cars per 24-hour day. Some of the engines were 720, 722 [pictured on the next page], 723, 725 and 727 and the engineers were George Knox, Joe Japp, Abe Coryell, McSwain and Llewellyn. McDougal was on the Phoenix passenger.

"Later Dan Fleming was on the Galloping Goose [the railway motorcar or gas-electric used in the 1920s and 1930s]. Moriarty, Vetter and Smith were on the Oroville–Spokane passenger, with Smith on the Oroville end. Each worked two days per week, that was a round trip of 257 miles [413.5 km]. Brisenham ran the coke train from Marcus to the Grand Forks smelter. Double headers would leave the smelter at 7:30 a.m. and be back by 2:30 p.m. At the smelter there were two Sherbrookes and four Davenports [small tank locomotives].

"You had to watch the water level getting low on hockey specials climbing the hill to Phoenix...

...continues next page

multi-millionaire Alexander Cochrane, who was also a director of the Northern Pacific; Payne Whitney, son of the United States Secretary of the Navy and one of the New York directors of the Granby Company and of the Northern Pacific Railway; Gardner M. Lane of Lee, Higginson & Co., bankers of New York and Boston; Samuel and Jonathan Horne, capitalists of New York; George C. Clark of Clark, Dodge & Co., also New York bankers; A. T. French, vice-president of the Manhattan Trust Co.; and New York bankers August Kissel and Fletcher A. Baker. Jay P. Graves and General Superintendent Hodges, as they had at the mines, gave their well-heeled guests a tour of the smelting works. "The party spent about an hour in the smelter, and at 3:00 p.m. the long train pulled out from the yards, with 'Jim' Hill on the rear platform waving a cheery good-bye and bound for the coast." If local reports are to be believed, the party was most impressed with the development of the mining industry in the region. "From remarks made," *The Phoenix Pioneer* concluded, "it was plain that they were more than pleased and satisfied, not only with the extent of the ore bodies, but with the manner in which the entire business of manufacturing copper is being carried on by the Granby Co."[20]

The domination of the party by representatives of American banking and railroad interests reflected recent changes in the ownership of the Granby company and the overall control of so much of the Boundary District's production by American capitalists. A syndicate led by William H. Nichols, "a transcendent figure in the copper industry of the United States," who as John Fahey notes, "held the center of the copper business; he was one of those who manipulated supply and price...," had bought large blocks of shares from Jay P. Graves and other investors. They soon replaced most of the Canadian investors and others on the board of directors.

The new Great Northern line to Phoenix gave Nichols a direct route to the United States and his

"Engineer, fireman Val, Conductor, two brakemen, call boy make up the crew of a work train here... Val," the fireman wrote on the back of this postcard of a Great Northern train on the Phoenix-to-Grand Forks smelter run. The engine is GN's 722, a well-maintained G-3 class 4-8-0 built by Rogers in 1899 and still on the GN roster until 1949. A photo of it at Kettle Falls in 1947, about 40 years later, appears on page 179. The only major changes after all those years were its conversion to burn oil, new cylinders and the addition of superheaters.

On the facing page, the daily mixed train steams over Bridge 66 with a refrigerator car behind the engine.

–Robert Turner Collection, left; Wilson Collection, Okananan Archives Trust, far left.

eastern refinery interests, and strengthened the position of Spokane in the commercial struggles for the business of the region."[21] Hill was a master player in these high-stakes business dealings behind railway construction. By 1905 he had purchased a substantial interest in Granby Consolidated and his representatives were soon on the board of Granby, giving the Great Northern substantial leverage over freight rates and the traffic of the Granby mines and the Grand Forks smelter. When coupled with his investments in the Crow's Nest Pass Coal Company, Hill and his associates were achieving an integration of the railway with the coal mines of the Crowsnest, the mines and smelters of the Boundary and the refineries in the east. It was a powerful combination soon reflected in the operations of the Boundary railways. It is also possible that the heavy investments at Phoenix and in Granby were to ensure success after the failure of traffic to materialize as hoped on the line to Republic.

In Vancouver a few days after his successful tour of Phoenix, Hill continued to promote the development of the VV&E across southern British Columbia. Construction work, he assured the newspapers, would begin that winter at Cloverdale, in the Fraser Valley, where it would join with the GN-controlled New Westminster Southern, and proceed to Hope and beyond into the mountains to connect with the other sections of the line

Bill Euerby Remembers ...

...as locomotives would pass over the summit at the Canadian Pacific Phoenix station and start downgrade on the Brooklyn Spur and could uncover the crown sheet if the water was low. [If the crown sheet in a heated boiler is exposed above the water level due to insufficient water it can lead to very rapid overheating and a catastrophic boiler failure, usually an explosion.] Hockey teams from Trail, Phoenix, Greenwood and Grand Forks all travelled on special trains for matches."

–Bill Euerby, interview with Dave Wilkie, c. 1970.

Great Northern engines steam quietly beside the No. 3 tunnel of the Granby Mine at Phoenix. Behind are the huge loading bins of the Old Ironsides Mine. On the facing page, two GN G-3 class 4-8-0s slog up the 3-percent grade at Mile 3.7 on the Phoenix branch with 20 empty steel "battleship" ore cars. The G-3s were rated at 340 tons each on this grade.
–J. McRae photo, Wilson Collection, Okanagan Archives Trust; BC Archives, 44087.

being built into the Similkameen District. "It will be an all-Canadian line, with terminals in Vancouver," Hill proclaimed in his usual emphatic and bombastic style. "Mark my word: We will do more towards building up Vancouver than any railway corporation has done yet."[22]

Meanwhile, as Hill postured and impressed the press, traffic boomed on the Great Northern's Phoenix branch, and the GN's advantageous position became clear over the next months. The reasons behind Canadian Pacific's successful manoeuvres to delay GN construction were evident in the changing traffic patterns as Great Northern trains carried more and more ore from the mines. Rates on Granby ore from Phoenix to their smelter dropped from $1.00 a ton to 25 cents a ton and for a time GN controlled a majority of the Granby traffic. As John Fahey notes, although Canadian Pacific had cut its rates to the minimum to retain a small profit, Hill's GN "at first hauled Granby ore consignments at cost. Consequently, the Great Northern between 1904 and 1910 would carry threefifths of the freight of Boundary and 68 percent of Granby's." He also notes that the GN raised its rates as traffic increased.[23] Geography also had an influence, and some mines were accessible only to one railway.

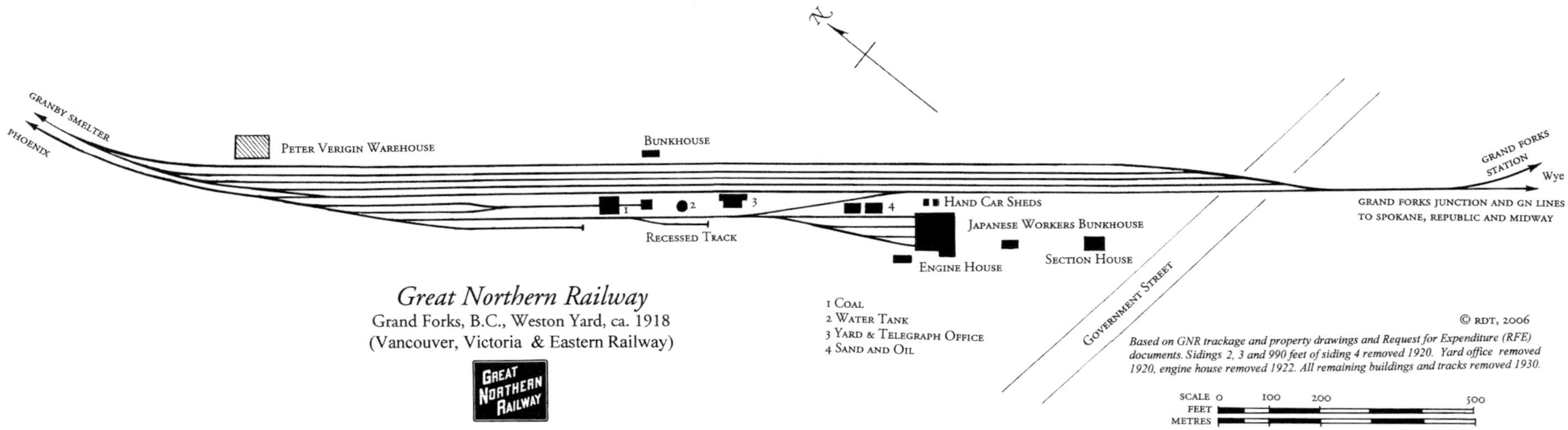

Great Northern's Grand Forks...

An overview of the Great Northern Weston yard to the west of Grand Forks with the CPR yards in the distance at left. This photo shows the coal dock, water tank and three-stall engine house before the fire of January 6, 1916, shown below. Another fire in December 1907 burned an earlier structure. A four-track engine house was built after the second fire. In the centre photo, GN workers pose in front of the building, and at right Tim Graham, Andy Kleman, Jack Green and others are beside the engine house.
–Dave Wilkie Collection above; below Boundary Museum, 991.055.206, 208 and 207.

The GNR's modest wooden depot at Grand Forks faced southeast at the corner of Donald and Indiana streets on a track from the wye at the south end of the yards. See also page 16. The depot was sold in 1930. A large nearby freight house was sold in 1924.
–BC Archives, 44215.

The CPR and Kettle Valley Union Depot in Grand Forks was an attractive red-painted structure near the city's centre. It was built by the Kettle Valley Lines after tracks were extended through the city centre in 1907 as part of the new line to Lynch Creek. The station burned in 1908 and was rebuilt. In 1913 it was enlarged as the Union Depot, trackage was expanded and connected with the CPR just north of its yards. Earlier, the Kettle Valley Lines' station, built in 1903, was south of the Kettle River near the CPR. CP trains started using the spacious Union Depot, with its large platform, on May 1, 1913.
–BC Archives, 44214.

Canadian Pacific's Grand Forks...
The CPR's Columbia & Western station of 1899 at Grand Forks was a typical wooden-framed structure with attached freight and baggage rooms and living quarters for the station agent. It survived the abandonment of the railway to continue serving the community. The CPR's substantial roundhouse at Grand Forks, built in 1912-1913 to replace the facilities at Eholt, is shown at left. A smaller, earlier structure was flattened by high winds in July 1905.
–Boundary Museum above, and at left, 995.049.017.

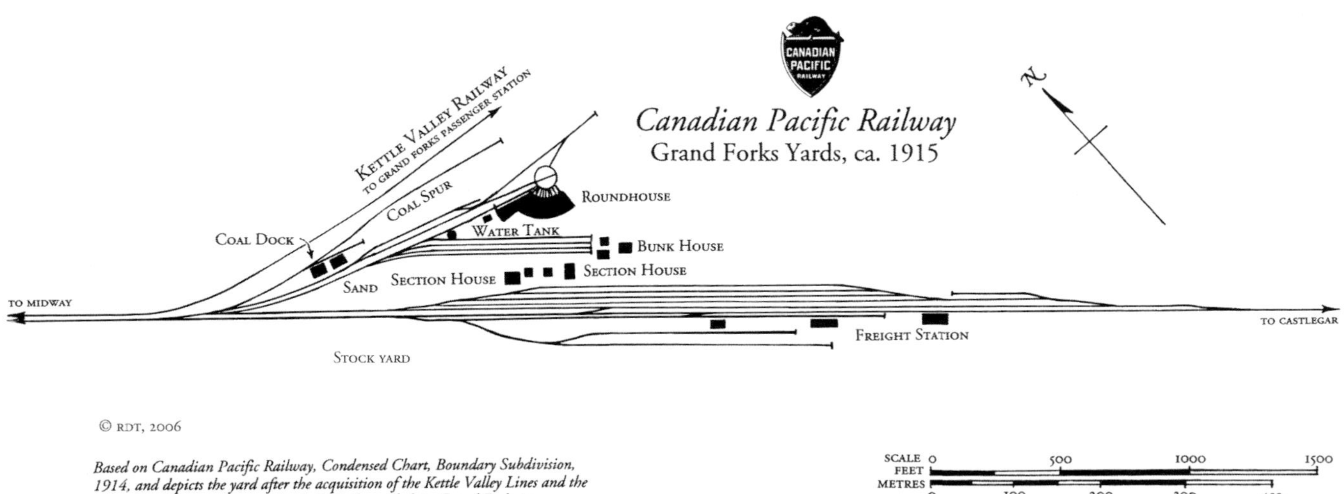

Based on Canadian Pacific Railway, Condensed Chart, Boundary Subdivision, 1914, and depicts the yard after the acquisition of the Kettle Valley Lines and the relocation of locomotive servicing facilities from Eholt to Grand Forks in 1912-1913.

Canadian Pacific's yards were to the northwest of the city, just south of the junction with the Kettle Valley Lines' tracks to the Union depot. The yards changed over the years as the KV Lines' trackage was integrated with the CPR and when CPR engine facilities were moved to Grand Forks from Eholt. Between 1913 and 1921 passenger trains took the KV Lines' tracks through the city to the Union Depot, but after 1921 when the bridge over the Kettle River south of town failed, breaking the route into the city, westbound trains backed down to the depot from the main yards and headed back out while eastbound trains did the opposite. This practice ended in 1952 when trains stopped only at the C&W station shown above. Later the downtown tracks, which had also served fruit packers, were removed.

This 1909 Phoenix line wreck, in the top photo, cost engineer Leo McAstocker his life. The second photo, taken on August 23, 1904, shows the wreck of CPR Shay 1901. With 24 loads of ore, one empty car and caboose, it was easing downgrade between Hartford Junction and B.C. Junction. Apparently an angle cock may have been closed from being hit by a chunk of ore, but in any event the brakes failed and the train ran away. The crew tried to hold the train with handbrakes and by reversing the engine, but the reversing "wiped out the teeth off all the crown gears on the wheels and tore her valve gear apart." The crew jumped and the train continued to "a curve near Denoro City before piling up in a heap and scattering 750 tons of ore over the landscape."[24] Surprisingly there were no serious injuries.

–Boundary Museum, 987-122-002.2; BC Archives, 37281.

Accidents and Tragedies on the CPR, GNR & Smelter Trains...

The steep grades and early braking systems on the heavy ore trains posed ongoing hazards to the train crews who worked to and from the mining camps and around the mines and smelters. The mining camp branch lines of the Boundary District and Republic were dangerous to work on. The references in the local newspapers to the "White Wing," a nickname for the local Canadian Pacific passenger train, none too subtly indicated the hazards of the line and the possibility of passengers and crew joining the angels. Indeed, there were more accidents there than on the well-known descent of the original CPR main line down the Field Hill in Kicking Horse Pass, where hazards have often been exaggerated. The following examples, and this is only a partial list, illustrate how dangerous and unpredictable railroading in the mountains could be, despite careful precautions and the best efforts of the railway crews to operate the trains safely and without incident.

On April 28, 1903, F. E. Donohoe, a 42-year-old brakeman, was riding on the footboard of the caboose of an ore train backing up the Phoenix Hill. He slipped and fell under the train, which rolled over his neck, killing him instantly. Engineer Peter Christenson, running a slag train at the Greenwood smelter, fell from a bridge while trying to re-rail his locomotives and died from his injuries in mid-April 1904. Just a few days later, three men were injured in a collision between a yard engine and four ore cars at the Granby Smelter at Grand Forks. The fireman was hurled through the cab window and knocked unconscious and the conductor was almost buried in ore spilled from the overturned ore cars.[25]

The winter of 1904-1905 produced a series of tragic fatal accidents on the trains. On December 13, 1904, D. K. McLean, brakeman on a GN construction train on the new Phoenix branch, was killed when several cars of the train ran over him, "scattering fragments of his body over the track for several yards." Apparently he was tightening a handbrake with a brake club when it broke and he fell between the cars. The same report carried the equally tragic story of CPR brakeman Dorst who also fell to his death while stepping from one car to another on an ore train between Eholt and Hodges siding. Dorst died on December 14.[26]

Great Northern conductor Michael Kelly, who a few weeks earlier had been on the construction train when brakeman McLean was killed, died instantly when his caboose and two flatcars derailed and rolled down a bank after hitting a handcar near Curlew. Kelly, a widower from Spokane, left two children and other relatives in Spokane. Just a few days later, on January 15, 1905, William Reeves, a GN brakeman, was killed on a construction train between Grand Forks and Phoenix when he fell under the train while trying to reach the engine.[27]

Fortunately, not all accidents had such terrible consequences. On the evening of January 14, 1907, the GN's Grand Forks-to-Phoenix passenger train hit an open switch near the Oro Denoro Mine where everyone onboard went for an unscheduled ride down to the CPR tracks below. As *The Phoenix Pioneer* reported, "...the Jim Hill train ran onto this spur down grade past the Emma Mine through other switches and finally derailed as it entered the CPR main Phoenix–Eholt line at B.C. Junction. The train consisting of engine and two coaches ran off the track tho' none was hurt."[28]

Unfortunately, there was little respite from tragedy. On February 2, 1907, a CPR train of 19 ore cars, one passenger car used on the mixed train and locomotive 1384 "was demolished [except for the coach] in the twinkling of an eye." Howard Rushton, brakeman, was caught in the wreck and had no chance of escape. *The Phoenix Pioneer* recounted the horrible sequence of events: "...the crew being the 'White Wings' crew, which does general work on the Hill when not on the passenger run....[Sandy] McGuire discovered his air did not work well then as a final resort an attempt was made to set the brakes more firmly by hand but it could not be done fast enough and the train was steadily gaining speed and was out of control. When it passed the Snowshoe mine it must have been travelling at a 50 mph [80 km/h] gait and at the Rawhide the momentum is thought to have easily exceeded 75 mph [120 km/h] on a 3 percent grade. Half a mile below the Rawhide mine the wreck occurred. Engineer McGuire stuck to his post till it was sure death to stay longer and with the fireman and one brakeman he jumped to safety having narrow escapes from death as it was. ...Howard Rushton was evidently trying to get down from an ore car to jump when he slipped and fell and was rolled over and over and ground to pieces by the flying ore train... ."[29]

There was no mention of passengers riding this particular train but if there were any, we can assume they escaped uninjured along with the conductor who was probably riding in the coach which did not derail.

More Accident Reports from The Phoenix Pioneer, and Grand Forks' Evening Sun...

Early in January 1907: Eleven loaded GN ore cars ran off the end of a switchback, and the "loco climbed a tree."

June 6, 1907: At "...about 8 o'clock, a Great Northern ore train was wrecked about five miles out of Phoenix at the first derail switch. No one being hurt."

August 22, 1907: "[The]... morning's passenger train on the GN leaving here at 9 o'clock was wrecked about seven miles out of Phoenix above Denoro, two brakemen were injured. Traffic was blocked for two days."

September 18, 1909: "Fallen tree sends ore train on mad plunge... a small tamarac tree five inches in diameter, blown across the Great Northern track on one of the switchbacks from the Granby Mines, was the cause of a serious and fatal wreck in Phoenix at 11:30 Thursday night and sent Fireman Roderick Olsen and Engineer John Lobner to their death. Both the fireman and engineer appear to have stuck to their posts, instead of jumping, and went down with the engine. ...Olsen was found under the wreck, having been killed instantly. Lobner was pinned under the cab, fatally scalded and injured, and asked his rescuers to end his sufferings. Dr. Dickson gave him temporary relief and he was quickly removed to the hospital in the Granby ambulance, death ending his suffering at 4:30 a.m. His wife arrived from Grand Forks at 6:30 on the special train conveying the wrecking crew."[30]

October 31, 1909: "While running light along the Phoenix branch of the CPR about 7 o'clock Sunday morning, locomotive 1385 suddenly left the rails on the big curve about a mile from Eholt and plunged over an embankment three or four hundred feet in height. The engine was manned by Engineer Leo McAstocker and Fireman George Beattie. Neither had time to jump. McAstocker had his skull badly fractured. Beattie was thrown into a tree, falling to the ground a little later. [At the Greenwood hospital], McAstocker died at 9 o'clock Monday evening without having regained consciousness from the time of the accident. His sister arrived in Greenwood Sunday evening on the Superintendent's special from Nelson.... Deceased was 21 years of age and was married a few months ago to Miss Luse of Eholt. George Beattie, the fireman is about 18 years of age. Mr. McWilliams who lives in a cabin near the place where the wreck occurred... made fast time to Eholt to report the accident. His prompt action probably saved another wreck. When the 1385 plunged over the hill it broke a rail and tore up a little track. Mr. McWilliam's prompt action enabled railwaymen to reach the scene and flag an ore train before it ran into the broken rail."[31]

Unfortunately, the dangers of railroading were not limited to the lines to Phoenix and there were other mishaps, wrecks and derailments along the Boundary railways. Some were caused by the weather, others by slides and natural hazards, some by human error. One attributed to a railwayman's mistake was on Farron Hill just after midnight on March 8, 1909, when engine 1381 exploded while taking water. The engine had just arrived on an eastbound freight from Eholt and was being prepared to continue on to Nelson. Headend brakeman Donald McQuarrie and fireman Cecil Glover, who was on the tender, died shortly after in hospital, despite being rushed to Nelson on a special train. The inquest was inconclusive, but the evidence and later enquiries suggested the engine's crown sheet had become dry, resulting in the explosion. Engineer Peter McCormick, despite his denials that water had been low in the boiler climbing Farron, was dismissed from company service.[32]

And so it went, with the crews in a constant struggle, day after day, week after week, to keep the traffic moving on time and safely with a rugged, unforgiving and rather primitive technology on mountain railways that brooked little or no oversight or mechanical failure.

GN brought in a steam wrecker from Spokane to clear up after a derailment in May 1905 two miles (3.2 km) below Hale siding on the Phoenix branch. The wrecked locomotive is 4-8-0 No. 723, and fortunately no one was injured. –Dave Wilkie Collection.

Reciprocity, the Great War and the Boundary Railways

Special Trains, Uncertain Operations, and an Uncertain Future...

"A special CPR train ran from Brilliant to Grand Forks carrying 100 Doukhobors to the land near the latter city lately purchased by Peter Verigin for his countrymen at a cost of $10,000."

—*The Phoenix Pioneer*, June 19, 1909.

"The GN coaches were crowded to their capacity on Wednesday morning. The heavy travel is due to the temporary suspension of the Granby Mines, most of the men will return after a brief visit to Spokane and other points. Granby workings to resume June 15."

—*The Phoenix Pioneer*, May 27, 1911.

"The smelter of the Granby company went down yesterday and will remain closed indefinitely... as a result of the war. The mines have also shut down.... The Greenwood smelter will go down Sept. 1st, if not before."

—*The Grand Forks Gazette*, August 3, 1914.

"Nearly half the population of Grand Forks was at the Union depot Thursday to give a hearty and loyal send-off to Grand Forks volunteers for active service in the present crisis.... At the depot citizens had turned out en mass to see them off, the band playing national aires. Several photographs were taken and after time had been given to say good-bys the train pulled out amid cheers form the throng and 'Auld Lang Syne' by the band."

—*The Grand Forks Gazette*, August 29, 1914.

James J. Hill had long been a strong advocate of reduced tariffs leading to free trade, often called "reciprocity," and increased commerce. He frequently spoke publicly on the subject to trade and industry audiences and his views, as a major business figure, were widely reported. Although free trade was always a contentious subject, he nonetheless had receptive audiences, and preached for cooperative, rather than competitive, trade between the two countries. Hill summarized his views in a 1908 interview, quoted in the adjacent sidebar.

There was of course opposition from many quarters including grain producers who worried about loss of markets and competition. Hill lobbied with politicians and businessmen and advised President Howard Taft on the issue. Finally, in 1911, free trade legislation passed Congress but it did not fare as well north of the border.[2]

In Canada free trade was opposed by the Canadian Pacific and the federal Conservative Party that had formed the government during the CPR's construction. Manufacturing interests feared they would be driven out of business by larger firms operating in the United States, and others simply feared domination of commerce by Americans. However, since 1896 a Liberal government under Wilfrid Laurier had been much more supportive of reciprocity between the two nations. The issue was far from simple in no small measure because import tariffs represented a major source of revenue for the Canadian federal government. Freer trade had been a part of Liberal policy, advocated with varying intensity, for the entire period of Great Northern expansion into southern British Columbia. As more and more Great Northern trackage was built back and forth across the international border, and it must be added Canadian Pacific took assertive measures of its own in services and through expansion, the GN was in an increasingly good position to benefit from the reduced trade barriers that

James J. Hill on Free Trade...

"It remains for business men in both countries to take hold of the subject before it shall be too late. No one can go into the matter, dismissing bias and self interest, without believing that the consummation most to be wished for is the wiping out of customs houses along our northern frontier altogether, and the establishment thereof of absolute free trade. It is the suggestion of natural law, of business interest, of the common good. If the time is not ripe for that, the least it demands is a policy and measure of ample reciprocity."[1]

–*The Daily Colonist*, August 25, 1908.

Photo, facing page...

The expansion of the Granby Smelter Works in its first decade produced a massive facility. The circular smokestack built in 1905, at left, was 150 feet (46 m) high and 13 feet (4 m) inside diameter. The flue chamber was 800 feet (244 m) long. Inadequate to permit further furnace expansion, in 1907 a rectangular steel flue chamber 13 x 15 feet (4 x 4.5 m) and 313 feet (95 m) long was added. By this time the surrounding hillsides had taken on the characteristics of a moonscape, nearly devoid of vegetation, and the foreground was rapidly filling with slag.

–Boundary Museum.

Special Instructions, Boundary Section...

"Engineers must use extra caution in approaching and passing places where Slides and Washouts are liable to occur.

"Freight trains must not exceed a speed of Twenty miles per hour when descending grades Farron to Cascade, Farron to Mile Post 33, and Eholt Jct. to Grand Forks.

"Ore Trains must not exceed a speed of Fifteen miles per hour Eholt Jct. to Granby Jct. and Farron to Mile Post 33."

"Regulations for Operating Freight and Ore Trains on Grades Over Two per cent on District Three....

"The normal position of a safety switch is for the spur and no engineer must whistle for switch to be set for main line unless speed is within required limit of eight miles per hour, and he is satisfied he can maintain it so. Engineers must promptly report any case of a switch being turned for main line before they signal for it. In case of failure to so report engineers will be held equally responsible with switchmen."

–Canadian Pacific Railway, *Pacific Division Time Table 15*, Sunday, June 6, 1909.

always seemed elusive. The year 1911 proved to be pivotal. With the passage of legislation in the United States, the negotiation of a free trade agreement for natural products and some manufactured goods, and a federal election looming in Canada, it looked as if free trade could become a reality, at least if Laurier were re-elected. But in the federal election of 1911, Wilfrid Laurier's government was defeated by the Conservatives led by Robert Borden, and Laurier resigned from power on October 7. Borden succeeded in mobilizing public opposition to free trade, particularly in the central Canada manufacturing centres, as well as capitalizing on other issues, including the establishment of a Canadian navy, and the free trade initiatives died. With that change in political fortunes, the steam went out of Hill's plans for expansion in Canada. Great Northern growth in B.C. all but ended, but some construction work continued to fulfill promises and expectations important to the image of the company and the reputations of James J. Hill and eventually his successors. In large measure, the field became open, unopposed for the CPR. A careful GN withdrawal began including trackage to the southern end of Kootenay Lake, the abandonment of the moribund Kaslo & Slocan Railway, and the curtailment of GN-owned sternwheeler service on Kootenay Lake under the Kootenay Railway & Navigation Company.

A year later, in 1912, the aging James J. Hill, by then 74 years old, retired as chairman of the board of the Great Northern and was succeeded by his son Louis. By the end of the fateful summer of 1914, the War dominated politics and government in Canada, leading the federal Conservatives to join forces with the Liberals to form the Unionist Party, with Borden as leader, between 1917 and 1920. With little likelihood of support in Canada, free trade initiatives on the American side sparked little interest.

Meanwhile, along the Boundary, the period leading up to the failure of reciprocity initiatives and the tumultuous years of the First World War brought many changes, one of the most important being the acquisition of the Kettle Valley Lines' Canadian interests by the Canadian Pacific. The CPR had developed a "close relationship," as John A. Eagle points out, with the Kettle River Valley Railway overseen by J. J. Warren at the time the Great Northern was building to Phoenix and Midway. The CPR had underwritten the construction of the line north from Grand Forks to Lynch Creek in 1905–1906, and in the next few years surveys were carried out between Midway and the coast. The CPR funded KRVR construction and surveys in exchange for bonds and debentures preparatory to resuming construction to the coast, as described in the next chapter.[3]

For the Kettle Valley Lines' trackage in Washington, known as the Spokane & British Columbia, a far less optimistic future developed. The times were tough indeed and it had clearly lost any hope of gaining an edge over the Great Northern in the ore business. Its June 1912 year-end review was terrible. With a ratio of operating expenses to income of 253, it showed the worst financial returns of any railroad in Washington State. Only $809.10 was received in passenger revenue for the 1,837 passengers carried, and freight brought in just $20,648.77. Overall, the financially crippled line carried just 706,751 tons of freight one mile, meaning that just 18,357 tons was carried the full 38.5-mile (62-km) length of the line. Perhaps even more revealing, that modest figure translated into an average of less than two 30-ton carloads of ore or freight a day. The operating ratio fell to 177 over the next two years as traffic improved somewhat, but the financial picture was still so bad it is a wonder any trains were run at all, as the railroad went deeper into debt.[4]

Railway operations were inevitably linked to the mining and smelting activities and the world markets for metals, coke and coal. A protracted miners' strike in the Crowsnest in 1911 brought much of the production in the Boundary District to a halt for about five months and it did not resume until almost the end of the year.[5] In the meantime, the CPR began work on a new bridge over the North Fork of the Kettle River on its line to the smelter. This steel structure, spanning 350 feet (107 m) and supported on masonry columns, was estimated to cost $100,000, and 60 men were employed at the site over the winter of 1911–1912. At the same time, the railway began work on establishing Grand Forks as a division point which included moving most of the facilities from Eholt to Grand Forks. Forty parcels of land were purchased for $8,000 for the new terminal.[6]

By mid-January 1912, after the lengthy curtailment and shutdown of mining and smelting due to the coal miners' strike in the Crowsnest the year before, production had returned to normal levels. Both railways were busy hauling ore and serving the needs of the smelters and local residents. Great Northern was operating two double-headed ore trains every second day and on alternate days a single engine handled the tonnage, moving an average of 38 cars every 24 hours. On the Canadian Pacific, a single ore train each day was moving 19 cars of ore to the smelter. Coke shipments from the Crowsnest, much of them coming over the Great Northern, arrived regularly and six of the eight blast furnaces at the Granby Smelter were in operation. At the smelter itself a new slag granulating system that had been installed the previous summer was being tested and when in full operation the next summer it would eliminate the need for the small locomotives and slag pots and about 16 jobs. By the end of January 1912, the two remaining idle furnaces were blown in, putting all eight into production, and copper prices were rising. The B.C. Copper Company's smelter at Greenwood was also in production treating ore from the Mother Lode, Rawhide, Athelstan, Emma and other mines. On February 3, *The Grand Forks Gazette* reported that 2,000,000 pounds (over 900 000 kg) of copper had been produced since January 1, and on March 9 the paper reported that Granby had a set a record by processing 25,156 tons of ore in the previous seven days. That summer the Republic mines were shipping 100 tons of ore per day to Trail for smelting.[7]

Ore supplies sometimes limited the smelting operations and as early as 1912 the Greenwood smelter began to operate below its production capacity. By that time too the Granby company was under no illusions about the long-term viability of its smelter at Grand Forks and of the Phoenix mines, diamond drilling efforts having almost fully explored the ore reserves at Phoenix. As a 1912 study of the district noted, "the mineralized zone of Phoenix has been well prospected and by such methods as render it impossible that any large body remains undiscovered. The main ore bodies of the camp have been located."[8] Granby had recognized that it would need to develop new ventures if it was to continue in the mining and smelting business and by this time had begun developing a new property at Hidden Creek on Hastings Arm on the coast just south of the border between Alaska and British Columbia. In this very remote site, known as Anyox and accessible only by sea, a townsite and smelter were built and a large copper mine was opened. In March 1914 the smelter was blown in and production started. The company also developed coal mines on Vancouver Island at Cassidy, south of Nanaimo, to supply the smelter, and it had other mining properties in Alaska. Anyox was to remain in operation until 1935 when the smelter closed and the property was sold to the Consolidated Mining & Smelting Company.

Despite the gloomy long-term outlook for the Boundary mines and smelters, production continued normally for some time and with it the operation of the railways. On May 5, 1912, fire destroyed the CPR roundhouse and damaged an engine at Eholt, but during the next month preliminary work started on a new 10-stall roundhouse and machine shops at Grand Forks which were said to cost as much as $180,000. The city contributed to the relocation by supplying water at a fixed rate, allowing a 10-year tax-exempt period and providing a free lot, within one half mile of the CPR depot, to CPR employees who owned a house and lot in Eholt. The little community at the junction to the line to Phoenix certainly seemed doomed, because at the end of August, another fire destroyed many of the town's business buildings. Of course it is possible that some business owners, knowing the divisional point was moving, may have been just as happy, at least if they had adequate insurance coverage.[9]

Construction on the new division point facilities was

Chinese Were Unwelcome...

"Contractors on the new [CPR] railway to Wellington Camp were indignant at the insinuation that any of them employed Chinamen. The [Chinese man] who landed in the local lock-up as a result of a fracas while celebrating his New Year was employed in a boarding car or camp connected with the CPR and not in any of the camps on the new railway. Mr. Tierney only employs white cooks."

–*The Phoenix Pioneer*, February 12, 1910.

Not Just Smelters; Fruit of the Fields...

"Sixty carloads of apples are already marketed. ...at least 100 carloads will be marketed before the season is over. Of the apples marketed, 49 carloads were shipped to outside points. The Grand Forks Fruitgrowers' Association have been the principal shippers, having sent out 30 cars, most... to prairie cities and one to Australia.

"Three carloads were sold to the Hudson's Bay Co. at Calgary, and the company [announced] 'These are, without exception, the finest lot of apples we've seen, and we doubt very much if old Mother Nature produces any better. And they are beauties. ...every box contains the most luscious fruit; all are choicest No. 1 Mackintosh Red Apples...' "

–*The Grand Forks Gazette*, November 2, 1912.

GREAT NORTHERN RAILWAY

TIME TABLE

THE COMFORTABLE WAY.

S. F. & N. RY.

Daily Leave	PHOENIX	Daily Arrive
8:35 a.m.	Spokane, Seattle, Everett, Bellingham Vancouver Victoria, and all Coast points......	6:45 p.m.
8:35 a.m.	Spokane, Fernie, Winnipeg, St. Paul, Minneapolis	6:45 p.m.
8:35 a.m.	Grand Forks, Republic, Curlew, Marcus and Ferry (Midway)	6:45 p.m.
8:35 a.m.	Northport, Rossland Nelson	6:45 p.m
8:35 a.m. Ex. Sun.	Kaslo, Sandon.......	6:45 p.m. Ex. Sun.

Connecting at Spokane with the famous

"ORIENTAL LIMITED"

2—DAILY OVERLAND TRAINS—

From Spokane for Winnipeg, St. Paul, Minneapolis, St. Louis Chicago and all points east.

For complete information, rates, berth reservations, etc., call on or address,

M. M. STEPHENS, Agent, Phoenix.

S. G. YERKES, A. G. P. A., Seattle

A Great Northern newspaper timetable from the early 1900s.

–Dave Wilkie Collection.

NELSON AND MIDWAY

863	11	861	Mls.	TABLE 119	12	862	
NOON	P.M.	A.M.		(Pacific Time)	P.M.	P.M.	
	* 9.00		0.0	Lv......NELSON ‖ ⊙ (C)......Ar	*11.15		
	10.20		25.7	Castlegar	9.50		
	10.30		27.4	West Robson	9.30		
			30.1	Westley			
	f10.43		32.4	Labarthe	f 9.17		
	f11.07		39.0	Shields	f 9.00		
	f11.22		43.6	Coykendahl	f 8.48		
	f11.47		50.2	Tunnel	f 8.29		
	12.20		58.0	Farron	8.08		
	f12.34		62.4	Paulson	f 7.47		
	f12.46		66.4	Coryell	f 7.32		
	f12.58		70.6	Lafferty	f 7.16		
	f 1.17		77.0	Fife	f 6.55		
	1.30		81.9	Cascade	6.33		
	f 1.33		83.0	Billings	f 6.29		
	f 1.43		87.2	Gilpin	f 6.21		
	1.56		93.5	Cuprum	6.08		
	2.00	d 8.30	94.8	Grand Forks ⊙ (C)	6.03	d 5.00	
	f 2.27	f 9.00	100.4	Fisherman	f 5.39	f	
	f 2.45	f 9.30	105.4	Hodges	f 5.24	f	
d12.00	* 3.05	d10.00	108.8	Ar......Eholt......Lv	* 5.10	d 4.00	
		d10.15	108.8	Lv......Eholt......Ar		a11.45	
		f10.30	111.0	Coltern		f11.35	
			113.7	Silica			
		f10.39	116.3	Hartford		f11.23	
		d10.55	118.5	Ar......PHOENIX (C)......Lv		d11.15	
d12.30	* 3.27		117.3	GREENWOOD (C)	* 4.40	d 3.20	
1.05	f 3.40		121.1	Boundary Falls	f 4.30	f 2.55	
d 1.30	* 4.00		126.6	Ar......MIDWAY (C)......Lv	* 4.15	d 2.30	
P.M.	A.M.	A.M.			P.M	P.M	

Nos. 11 and 12 Standard Sleeper and Cafe Parlor Car Nelson and Vancouver.

–Canadian Pacific *Western Lines, Time Tables*, August 1, 1919, Robert Turner Collection.

Passenger Services to Phoenix Were Great Fodder for Grumblings...

On the Great Northern...

"Phoenix does not ask the Great Northern to run a palace parlour car to this city. Local multi-millionaires prefer to travel without aristocratic airs. But when the Great Northern expects fares for first-class trips out of Phoenix we demand there should be some effort to give reasonable service. A coach and smoker are pulled into Phoenix daily and might be all right if kept clean but when [crowding in] the smoker becomes such as to force smoking in the so called first-class coach, and that to say nothing of the cinder topped seats, in turn forces ladies to the rear platform, the accommodation becomes sheer contempt on the part of the railway company."

–*The Phoenix Pioneer*, December 25, 1909.

"For the past three months the GN passenger train has been arriving at Phoenix from half an hour to two hours late, and the delays have been occasioned largely by the company "doubling up" on its train cars and bringing ore cars up the hill with passenger and express coaches. The train usually leaves Grand Forks on schedule, but loses the time coming up the grade. ...This mixed passenger and ore car service is an injustice and... should not be tolerated."

–*The Phoenix Pioneer*, December 17, 1910.

"The Great Northern passenger train arrived two hours late Tuesday evening. This was due to the ancient type of locomotive in use being unable to navigate the grade with a coach and baggage car, taking four hours to go a distance of 20 miles." [This was likely former Spokane Falls & Northern engine No. 5 or 6.]

–*The Phoenix Pioneer*, October 23, 1909.

"A passenger coach fresh from the paint and varnish shop is now on the local run of the GN and is a vast improvement over its smoky predecessor."

–*The Phoenix Pioneer*, February 4, 1911.

Meanwhile, on the Canadian Pacific...

"The Canadian Pacific have recently put a unique carry-all coach on the Phoenix branch. It is a novel piece of rolling stock and its comprehensive design places the parlor-sleeping and... coach among the relics of ancient days. A scrutiny does not reveal the name of the designer, some credit it to Sir Thomas Shaughnessy's wonderful brain of ingenuity while others maintain it is the combined efforts of many noted designers. ...it is not unlike their every day coaches in outward appearance...[but] there are three important sections; one section has six upholstered seats and will carry as many as twelve passengers on a rush day. A wash-dish is located at the end of one of these seats and if there were water and towel in connection would be used quite frequently. There is a swinging door to the next section, where as many as eight people could be situated if room could be found for their feet. This is the "smoking parlor" and when the smoke gets thick it pushes the swinging door and circulates among the coach seat passengers. These two sections take up about a third of the coach, the balance is devoted to the conveyance of mail, baggage, express, freight, including both live and dead stock, and provides office accommodations, dining room, kitchen, parlour and sleeping quarters for the officials of the company. Sir Thomas has not been among any of the officials who have enjoyed this coach novelty as yet."

–*The Phoenix Pioneer*, December 31, 1910.

"The versatile accommodation freight and passenger coach has been retired from the local branch of the CPR and is replaced by passenger and baggage coaches more in keeping with the business offered by the company over this line. The new rolling stock, as long as kept clean, will make first-class equipment."

–*The Phoenix Pioneer*, February 4, 1911.

slow, but by early January 1913, the concrete turntable foundation was ready, work on the first five stalls of the roundhouse was progressing, and the machine shop and coal dock were complete. Trackage through the city was also being extended under Kettle Valley auspices, and on Third Street the Kettle Valley's station was expanded and opened as a new "Union" station, as it was called in the newspapers, to serve CPR and Kettle Valley trains under an agreement with the city providing tax concessions. Kettle Valley tracks were also extended around the city as part of its North Fork branch line construction. Soon after, the CPR moved its crews and their families from Eholt on a special 32-car train. For any families without housing at Grand Forks, the CPR supplied boxcars, probably actually work train accommodation cars, as temporary quarters, which in January probably provided only cold, cramped comfort.[10]

With the outbreak of the First World War in Europe in early August 1914, the Granby Smelter closed briefly because Germany had been a big purchaser of the copper produced at Grand Forks and it took a while for the demands of Britain and its Allies to take up the supplies of copper. Soon after, CPR crews, and presumably GN crews too, were feeling the effects as train service was reduced and men bid on jobs in other divisions. Passenger services were cut to three times a week to the Boundary District and similar service was provided to Phoenix. By early December the realities and complexities of the growing war were becoming evident and Granby and the mines resumed production as demands for copper and prices began to rise. Copper was a key strategic metal needed in munitions, communications and many other components of the expanding war effort. For the remaining years of the First World War, production problems most often related to maintaining supplies of ore, coal and coke to meet the enormous wartime needs for copper and precious metals.

The Greenwood smelter, which had also closed, reopened after 10 months in mid-June 1915, as did the Mother Lode and Lone Star mines. By early October, the situation had changed dramatically and *The Grand Forks Gazette* reported that Granby was breaking production records, with 4,000,000 pounds (over 1 800 000 kg) of copper produced in August, and B.C. Copper's Greenwood smelter also was busy, but suffering from a shortage of ore. Soon after, passenger service to the Boundary District returned to a daily schedule. During these difficult years there were many ups and downs in production fuelled by wartime demands, but disrupted by insufficient supplies of ore or coke, labour shortages and disputes. The difficult conditions continued even after the Armistice on November 11, 1918, when the catastrophic Spanish flu epidemic took its toll of workers and their families at Grand Forks and also in the Crowsnest Pass, and caused the smelter to close briefly because of a shortage of ore and coke.[11] The next few years were to be critical ones for the railways and mines in the Boundary District, but before that story unfolds, we must return to the pre-war expansion of the railways to the west toward the Similkameen mining and farming districts and the snow-draped mountains blocking the routes through to the coast.

Wartime brought uncertainty and then intense activity at the Boundary mines and smelters. Demands for resources grew, but so too did the needs for coal and coke and shortages sometimes limited smelting. The war drew manpower from the communities and many young men volunteered. The so-called "Great War" took its toll as the war memorials in the mining communities attest. This scene at Grand Forks' Union Depot on August 27, 1914 shows a special train with a large crowd in attendance giving the first 15 volunteers of the "The Grand Forks Sharpshooters" a warm and patriotic send-off, less than a month after the beginning of the war. It was a poignant farewell for everyone; it marked a time to which there was no returning.

–Boundary Museum 986.231.001.3.

PART 2: RAILWAY DREAMS AND REALITIES

Building Westward to the Similkameen

Construction near Myers Creek and Myncaster...

"When completed the tunnel will be some 800 feet [244 m] in length and will represent one of the most difficult portions of the road.... Another engineering feat of great importance was the erection of the long semi-circular trestle which takes the road across the valley at Myer's Creek, approaching within 20 feet of the American boundary before the curve turns to the basin of Rock Creek. This will be one of the prettiest and most thrilling sights on the road."

—*The Boundary Creek Times*, MARCH 9, 1906.

Endless Rumours and Much Hot Air...

"J. J. Warren, General Manager of the Kettle Valley Lines is at present in Grand Forks. His company intends extending its lines from Republic to Spokane. This announcement lends colour to the rumour that the CPR is backing the enterprise with a view to securing a valuable feeder in the North Fork branch and by making connection between Republic and Spokane securing a better route for hauling coal and coke from the Crows Nest coal fields to the Boundary smelters."

—*The Boundary Creek Times*, AUGUST 3, 1906.

After the War, a Brief Promise of Traffic...

"For some time past the regular tri-weekly mixed train on the Oroville–Princeton branch of the Great Northern has been entirely inadequate to handle the traffic... and for the past two weeks, alternate days, Sunday included has witnessed a special train leave Oroville with a string of empties and return with a load of coal. The Coalmont Colliery... are now in a position to supply... fifteen cars per day."

—*The Penticton Herald*, OCTOBER 18, 1922.

AFTER THE CPR REACHED MIDWAY IN 1900, ITS CONSTRUCTION BETWEEN THE BOUNDARY District and the coast came to a halt for some time. Having built what amounted to a new railway nearly all the way from southern Alberta through the Rockies, the Monashees and the southern mountain ranges of British Columbia, as well as buying the steamer line serving the region, the CPR needed time to regroup and refinance. Initially traffic was slow to develop and begin to repay the massive investment in the new construction. Moreover, the CPR had other demands on its investment capital on the Prairies, on Vancouver Island and elsewhere, including expansion of the smelter at Trail, the development of the Sullivan Mine at Kimberley in southeastern British Columbia, the construction of the famous Lethbridge Viaduct and, in the Rockies, the Spiral Tunnels, to improve its main line. The railway was undergoing major expansions and modernization as traffic increased

The VV&E's serpentine route west from Curlew to Oroville crossed the British Columbia-Washington border twice to maintain some semblance of a good grade, but even it was a difficult route. On the British Columbia side the line made a big loop at the small agricultural community of Myncaster where a substantial curved trestle was needed. It passed just 25 feet (7.5 m) north of the border before swinging away again to lead to Bridesville. The 2,420-foot (738-m) Chesaw spur crossed the border just west of the bridge to reach a grain elevator, but fell considerably short of actually reaching Chesaw. In the distance in the winter scene is Myncaster itself and the railway station. Even after the ends of the trestle were filled, it still completed almost a quarter circle on an 8.5 degree curve. The impressive trestle was 818 feet (250 m) long. Had the VV&E survived longer it is likely that the big bridge would have been replaced by a fill as was done for some of the approaches to the trestle.

–RON HODGE COLLECTION, AT LEFT; SUSAN DAHLO COLLECTION ABOVE, COURTESY HENRY EWERT.

The country station at Myncaster was typical of small depots along the VV&E and offered Railway Express, telegraph and passenger services. It was a fitting background for a family snapshot as was the section house nearby. The overview of the big trestle shows, looking north from the Washington side of the bridge, the *Okanogan Owl* on the span.
–Susan Dahlo Collection above and centre, courtesy Henry Ewert; Dave Wilkie Collection, bottom.

during a period of economic expansion and massive immigration to western Canada, and its investments in the Spokane International Railway, as noted earlier, were an important counter move to the Great Northern.

Complicating the construction of new trackage in British Columbia in the very early 1900s, the political climate in B.C. was unstable until the election of Richard McBride's Conservative government in 1903. McBride, who served as premier until 1915, was highly sympathetic to railway expansion, and was to make it a keystone of his successful political career and a prominent feature of his election platforms. Although it was several years before construction resumed on the railway west of Midway, the CPR began negotiations with the province over subsidies and land grant proposals for the completion of a Coast-to-Kootenay route. The narrative of Canadian Pacific's route from Midway to Vancouver are essentially the history of the Kettle Valley Railway and that complex story has been well told in a number of books.[1] In summary, Canadian Pacific gained control of the Kettle River Valley Railway and used it and its charter—officially renamed the Kettle Valley Railway following the passage of federal legislation in 1911—as its means of completing its line to Hope at the head of the Fraser Valley. The CPR also agreed to assume the financial obligations of the short-lived Midway & Vernon Railway, a railway proposal developed by local interests in the Greenwood area whose backers had completed little beyond grading a short section of roadbed just west of Midway. In June 1913, the CPR formally leased the Kettle Valley for 999 years. Far to the west, the KVR connected with the CPR's Nicola branch (more formally the Nicola, Kamloops & Similkameen Coal & Railway Company) at Merritt, which gave the proposed route to the Kootenays a direct link with the CPR main line at Spences Bridge on the Thompson River west of Kamloops.

The Kettle Valley Railway was managed for Canadian Pacific by James John (J. J.) Warren, who had been running the affairs of the KRVR in its fledgling days at Grand Forks, while Andrew McCulloch was the new chief engineer, chosen by CPR President Shaughnessy. McCulloch was later to win fame for the construction of the railway around the spectacular Myra Canyon high on the mountainsides above Kelowna and through the Coquihalla Pass in the Cascade or Coast Range east of the Fraser River. Interestingly, a few years before as a young engineer he had worked on the Columbia & Western construction. Federal legislation, passed in 1910, granted a subsidy of $6,400 per mile (about $3,975 a km) for up to 250 miles (just over 400 km) of railway between Midway and Merritt and a further similar subsidy for construction between Brodie (near Brookmere) and Hope. Provincial legislation, part of Premier McBride's railway policy upon which he had won re-election in 1909, granted further aid in the amount of $5,000 a mile ($3,100 a km) in cash or 3 percent bonds, for up to 150 miles (241 km) between Penticton and Merritt (Nicola), and a 10-year tax-exempt period for the remaining section of railway between Midway and Penticton. For these contributions, the CPR agreed not to hire any Chinese, Japanese or South Asian labourers on the construction, a position that was then, it is unfortunately necessary to note, politically and socially very acceptable. Later, in 1912, the province increased the subsidy with a grant of $10,000 per mile ($6,200 a km) for the very difficult work through the Coquihalla Pass and canyon and $200,000 for the Fraser River Bridge at Hope. The federal government added $250,000 for the bridge in 1912, and reduced the length of construction subject to subsidy to 200 miles (322 km).[2]

The Canadian Pacific did not wait for all of these arrangements and subsidies to be formally in place before work started. The railway had authority to construct a line with the charters it had acquired, and a great deal of survey work was already completed. The CPR chose a route that continued west from Midway to Rock Creek, swung north and crawled over the Okanagan Highlands, and wound along the cliff faces of Myra Canyon high above Kelowna before descending to Penticton in the Okanagan Valley where a division point was established. From there the tracks climbed again and ascended the western face of the valley, running north to Summerland before swinging west to climb the forested highlands and then descend to Princeton. There, the tracks were to proceed northwestward through Tulameen to Brookmere, swing generally southwest to ascend the Coquihalla Pass, and then follow the Coquihalla River to Hope before crossing the Fraser River to join the main line of the CPR on its north bank. At a junction called Brodie, the tracks also ran north to Merritt and a connection north to the CPR main line at Spences Bridge. The route sounds straightforward, until viewed on a detailed map, or as surveyors found it in the field, where mountains and swift rivers seemed to thwart their every plan. It was a difficult route at best, and an expensive one to construct and operate. For the men who would run the trains and maintain the tracks it was also dangerous and a seemingly never-ending challenge.

In the meantime, the Great Northern was not waiting for its rival. By 1905, it was pushing westward from Curlew, on the railway between the Republic and Grand Forks mining centres, towards Midway and beyond to the west. Hill had several objectives in mind including the completion of his much-touted Coast–to–Kootenay railway route. More immediate was gaining the potentially rich traffic of the Okanagan and eventually shipments to and from the mining camps of Hedley and Princeton in the mountains to the west in British Columbia. In Washington the Okanogan Valley was an area of potentially rich agricultural production that he was very interested in promoting and developing, and in British Columbia the town of Hedley, which will be described shortly, was developing as one of the richest gold mining claims in Canada.

Hedley, situated in the Similkameen Valley where Twenty Mile Creek[3] (later Hedley Creek) tumbles out of the mountains to join the Similkameen River, about 40 miles (65 km) to the east of Princeton, was not a roaring boom town like Phoenix or so many others in the west but one that had grown more slowly. "At no time in the 10 or 12 years of its existence," noted Charles Camsell in his report to the Geological Survey of Canada in 1910, "has it been troubled by a boom; a feature so characteristic of many modern mining camps, and, as a rule, so detrimental to the best interests of the mining industry."[4] Placer miners had worked the bars and gravel beds of the Similkameen River as early as the 1860s but the first mineral claims at what became Camp Hedley were not staked until 1894. These claims eventually lapsed but others, the Horsefly, Sunnyside, Copperfield and the Nickel Plate, were staked in 1897 by two Swedish prospectors named Wollaston and Arundel. They took ore samples—mostly arsenopyrite containing gold—to the Provincial Fair in New Westminster and there they attracted the interest of M. K. Rodgers, representing Marcus Daly, a mining and smelting magnate from Montana, a founder of Anaconda Copper and an important business associate of James J. Hill.[5]

Beyond question, the involvement of Daly in the early days of the Hedley mining venture, and the important role of Daly's Montana properties as major traffic sources for the Great Northern, influenced Hill's decision to build westward along the Similkameen. Compared to the Montana properties, the Hedley mine was probably a speculative venture for both Hill and Daly, but it held promise of lucrative returns in a newly developing territory where there was no competition as yet from other railways. To be successful in mining and railroads, it paid to look ahead and secure one's options, and, moreover, it complemented perfectly the expansion of the VV&E between Midway and Vancouver.

After further exploration by Rodgers, the claims were soon taken over by Daly and development work began in earnest in 1899. Daly's mines were situated high above the Similkameen Valley near the top of Nickel Plate Mountain, 4,300 feet (1300 m) above the town of Hedley. At that time, access to Hedley was difficult at best with all supplies having to come in over 28 miles (45 km) of rugged wagon road from Penticton, in the south Okanagan. Although Daly died of diabetes in Montana in 1900, his successors developed the properties to a high standard and, in the spring of 1904, a large modern mill was completed at Hedley in the valley below the mine.

Eastward over the VV&E...

"The Myers Creek valley runs north and south and crosses the international boundary, but the main line of the railway does not cross the line here. It runs along either side of the valley in the shape of a long condensed letter U or ox-bow, the curved base of which comes within 20 feet of the boundary line, where it crosses the valley on a high trestle.

"The station by which Myers Creek valley is served is Myncaster, called no doubt after McMynn brothers, who were pioneers of the valley and owned the principal ranch in it. From Myncaster a spur has been constructed across the boundary in the direction of Chesaw, but as yet no steel has been laid on it.

"Myers Creek was at one time an excellent trout stream and from all appearance it should still be so. The creek narrows into a canyon at two points between the McMynn property and ... the Kettle river, and between these points lies a beautiful stretch of meadow land formerly known as Nicholson's meadow, but now owned by F. Richter. A tunnel 900 feet [275 m] long emerges out on the Kettle river, within sight of Midway."

–*The Hedley Gazette,* July 23, 1908.

The Daly Reduction plant at Hedley.
–Mines Branch Photo.

Tunnels on the VV&E, Grand Forks to Keremeos.

1. Grand Forks–Curlew:
113 feet (34 m), 2.3 miles (3.7 km) east of Curlew, east bank of Kettle.

2. Midway–Molson:
900 feet (274 m), 6.2 miles (10 km) west of Midway near Myers Creek.

3. Midway–Molson:
116 feet (35 m), .75 miles (1.2 km) west of Bergen at Lake Bergen,

4. Midway–Molson:
113 feet (34 m), 1.9 miles (3.1 km) west of Bergen.

5. Midway–Molson:
350 feet (106 m), 1.4 miles (2.25 km) west of Myncaster, near Myers Lake.

6. Molson–Oroville:
448 feet (137 m), 1.3 miles (2.0 km) east of Oroville.

7. Oroville–Keremeos:
1,761 feet (537 m), 4.95 (8.0 km) miles west of Oroville.

–Sources: Raymond Corley and GNR survey maps, Dave Wilkie Collection.

Operated as the Yale Mining Company, the two major mines, the Nickel Plate and the Sunnyside, were highly valuable properties. By the end of 1905, when the Great Northern started westward from Midway, over 40,000 tons of ore had been treated.

In 1905, Hedley and Princeton remained distant objectives of the VV&E but at least construction west from Midway had begun amid promises of a rapid completion of the railway. Princeton, the then remote settlement at the confluence of the Tulameen and Similkameen rivers, had become an important distribution centre for the surrounding ranching and mining districts. Coal deposits, although a lignite not suitable for coking, and timber in the area held promise of development. Further to the west in the Tulameen Valley were even more encouraging coal deposits. However, any large-scale work awaited the construction of railways to provide the necessary transportation facilities.

Contractors Caughren & Waldren won the major contract for the next section in the VV&E's railway to the west. The route was complex: it left the Republic line at Curlew and followed the Kettle River in a northwesterly direction to cross the border at Ferry, Washington, just south of Midway, B.C. Whereas the CPR had gone north through Eholt and Greenwood on a mountain-climbing route that skirted the Phoenix ore bodies, the GNR went south, reaching Midway on November 28, 1905, on a water-level route that avoided any significant grades. They came to within a few hundred yards of each other again at Midway, then the end of track of the Columbia & Western. As if to demonstrate the alternatives available, the two routes were not to converge again until reaching Princeton, far to the west.

Great Northern crews carried on west of Midway and as many as 1,200 men were employed on the grading, tunnelling and track-laying work on this 30 miles (48 km) of railway. The principal contractor was John J. Stewart, and the manager of construction was Patrick Welch and many subcontractors were employed. Meanwhile, work was underway on the Washington side of the border, where the railway progressed over a distance of 47 miles (over 75 km) heading through Molson and down a series of sweeping curves to the Okanogan Valley.[6]

Meanwhile, there was a continued hiatus of another five years in CPR construction west of Midway after the GN tracks reached the little town. Before Canadian Pacific determined to begin construction again into what it saw as rugged, unproductive territory, the Great Northern's continued westward expansion was making substantial headway as it climbed over the Okanagan ranges, but the CPR did not resume construction until June 30, 1910, at first using some of the old grade of the Midway & Vernon Railway towards Rock Creek. The GN's westward push from Midway in 1905 precipitated another of the bellicose and rather stupid encounters between construction crews in the area. The melodramatic events are described in the sidebar on page 140. Despite the delays, the VV&E was pushed westward with two shifts working on grading and track construction that winter. After leaving the valley lands around Midway, the GN began a steady climb following Myers Creek to the west. The line required four tunnels and several substantial bridges. South of the present-day community of Rock Creek, the GN swung south along Myers Creek and then curved back north on an impressive loop over a trestle 818 feet (249 m) long and 90 feet (27 m) high. The trestle almost brushed the border just before reaching a station named Myncaster. "This [trestle] will be one of the prettiest and most thrilling sights on the road," commented *The Boundary Creek Times* on March 9, 1906. From there the tracks hugged the hillsides along the south side of Rock Creek, passed through another tunnel and followed the course of Baker Creek towards Bridesville. To gain elevation, the tracks made a double loop through the rolling hills near Bridesville, and then crossed the border before reaching the summit of the grade at Molson. There a small yard with water tower, wye and depot was located. The route west from Molson was all downhill over a seemingly endless series of loops down into the Okanogan Valley at Oroville.

The descent into the Okanogan Valley from Molson was spectacular. Molson, at an elevation of 3,706 feet

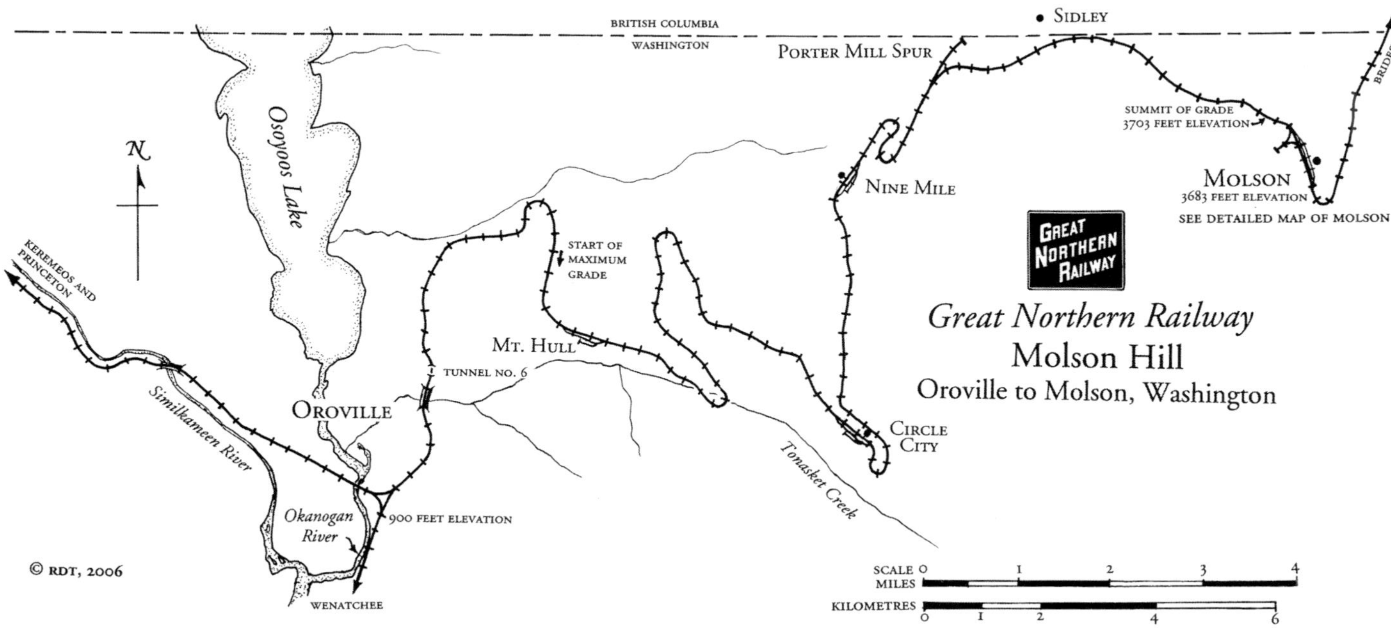

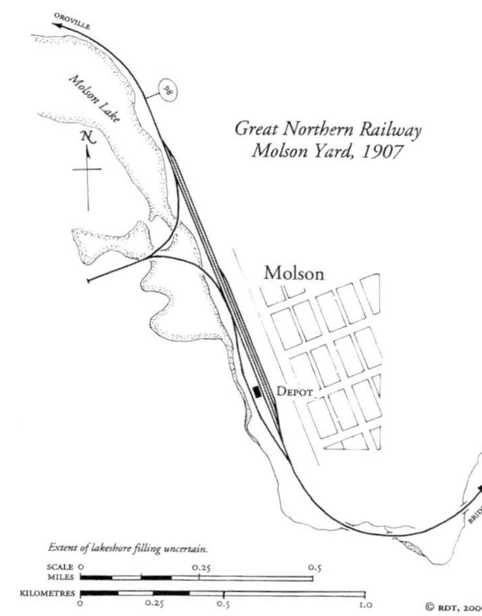

(1130 m), was the highest station in Washington, yet Oroville, on the Columbia River, had an elevation of only 926 feet (282 m). In a straight line the distance between the towns was just nine miles (15 km) but the trackage had to snake its way along the mountainsides for 27 miles (43 km) to maintain a still very steep grade of 2.5 percent down what became known as Molson Hill. The curvature of the tracks, if laid end to end, would have described nine complete circles. The one saving grace of the line was that the mountain slopes on the east side of the Okanogan, down which the rails twisted in their descent, were not in a heavy snowbelt.[7] The excessive grades and curvature were certainly not the standard of construction that James J. Hill usually achieved for his railroads. Indeed, they were a compromise at best, contributing to the early abandonment of the route.

Construction work on the line, described by the local press as "possibly the most difficult piece of railway building in the west," advanced rapidly during the winter and spring of 1905–1906. Two shifts were at work on grading and tunnelling with a large force of men being employed by the contractors. Tragedy hit late in February when two construction workers were killed. Then on Monday, April 16, 1906, a "missed hole" (a charge that had not ignited when a number of drill holes packed with explosives had been detonated) exploded in a rock cut being excavated 3.18 miles (5.12 km) west of Midway. At the time, a crew of labourers was in the cut and five Italians and one African-American were killed instantly. Six other Italian workers were seriously injured.[8]

As grading advanced, track-laying began west of Midway on July 24, 1906, with a crew of 75 men working on the construction train. By the beginning of October, the track crews had spiked down rails as far as Bridesville (then called McBride's). Despite a shortage of labour, the end of track was moving west at about one mile (1.6 km) a day. The October 26 edition of *The Boundary Creek Times* reported that Molson had been

Midway to Princeton...

The GN proposed a railway between Penticton and Oroville, but although surveys were made amidst press speculation, it was never built. It was for a time considered part of the railway from Midway as this quote suggests. "The section between Midway and Princeton will have a maximum grade of 1%, and will be very heavy work. The most interesting feature of this section is, that although the distance between Midway and the Okanogan River is only 32 miles in a direct line, it requires about 100 miles of a 1% grade to over come the summit, which is crossed at an elevation of 3,700 feet above sea level."
–*The Railway & Shipping World,* December 1901, p. 359.

The incomplete Midway & Vernon Railway grade near Midway hardly seemed worth fighting over, but it was really the charter and route that was important to the companies and their contractors. The original promoters of the line included James Kerr and Robert Wood of Greenwood.

–ANDREW MCCULLOCH COLLECTION, COURTESY BARRIE SANFORD.

Normally Quiet Midway…

"It would be difficult to find a place as much spoken of as Midway in relation to railway matters, with as little railway activity. Nevertheless it is today in possession of two rival railway systems, and there a competing point, but when there you would never think so unless someone whispered in your ear.... Midway is the home of blighted hopes and neglected opportunities...."

–*The Hedley Gazette,* JULY 23, 1908.

Midway's Merry Melee Melodrama ...

The Battle of Midway began in the autumn of 1905 when CPR construction workers placed fencing across the VV&E's already constructed grade about six miles (10 km) west of Midway on Block 2703, a section of land that had been previously granted to the C&W (Canadian Pacific) even though it was well beyond the end of CPR track at Midway. VV&E contractors came into conflict with the CPR when they began dumping rock and rubble from one of the tunnels onto CPR property. Evidently the CPR had a free hand in selecting blocks of property and this one covered land south of the confluence of Myers Creek and the Kettle River to the Washington border, and was an obvious attempt to block the VV&E's progress.

VV&E workers pulled down the fencing complete with "No Trespassing" signs. In counterpoint they found CPR men had laid logs across the fill east of Tunnel 2 and the construction railway had been lifted and the rails twisted around trees. Moreover, the area was guarded by as many as 50 armed CPR men. GN crews pulled back and their expropriation order was finally granted on November 4. Even then the CPR refused to withdraw, claiming the order was invalid and referred to different lands. In response on November 7, the GN brought in over a hundred men from nearby construction camps who descended on the disputed section. The vastly outnumbered CPR men beat a retreat. Such foolishness garnered an equally foolish response. On November 9 things heated up; the CPR managed to gather a force said to include 300 men armed with picks, axes and shovels. At this point the Midway police arrested F. W. McLaine, the CPR land agent, and the construction foreman, who they thought were responsible for the attack on the VV&E men. The VV&E men merely retreated a few hundred feet across the border. The Midway police released the two men later that night after extracting a promise not to make trouble. VV&E men quickly established Hooligan's siding on a flat spot a short distance east of the disputed site so that VV&E workers could be housed there in boarding cars.[9]

On November 10, Hill's forces received a special trainload of supplies, including a boxcar full of barbed wire, and reinforcements. The wire was hastily strung along the right-of-way, while CPR men tried to tear up the VV&E tracks. Fortunately, things calmed down and the next morning the CPR's men headed back to Midway and its saloons, only to find them closed by the police. Not deterred, the CPR men walked the half mile south to Ferry, Washington, where the saloons were filling up with VV&E men. The workers simply pushed their way in and sat shoulder to shoulder and drank, but one more round of foolishness remained. Three VV&E men went back to Midway and kidnapped McLaine and held him in a powder house near Tunnel 2. When the VV&E's chief construction engineer, James H. Kennedy, found out, he personally freed McLaine and made the men apologize. But the three kidnappers laid charges against McLaine, presumably for the earlier attack, and he countered with a charge of kidnapping. The police held all four in the local jail and by next morning they agreed to drop all charges. A November 11 report noted that each railway had about 150 men near the site, with each receiving $5.00 a day just to be there, a situation that none of the contractors appreciated. After that, the companies' lawyers took over and the dispute was resolved in court. The GN's original expropriation order had indeed referred to other property, but eventually all was settled and the gap in the VV&E closed, with legal process at last triumphing over stupidity. The workers probably enjoyed the entire fracas and felt better for it. The saloon keepers undoubtedly did.

While all this was going on, the CPR was also contesting right-of-way issues with the Midway & Vernon Railway, a fledgling attempt to claim a charter right-of-way after some grading, if not actually a completed railway, west of Midway. A report of October 27, 1905, in *The Boundary Creek Times* noted that the CPR had begun grading west of Midway to a point where the CPR thought the M&V right-of-way would cross their own surveys. By being first on the ground the CPR could force the M&V to cross the CPR and pay the costs. However, the CPR crews were thwarted when they found out the M&V had already purchased the property.

None of these events that swirled around the construction camps and courts had any great significance to the long-term future of the railways, but they did add spice to the lives of the track construction workers. Eventually the CPR acquired the charter of the Midway & Vernon, which was probably just what the promoters intended in the first place.[10]

reached and that the yard tracks and a wye were being put in. On November 6, 1906, train service was extended to Molson. From Molson west to Oroville down the steep face of the mountains to the Okanogan River, the pace of construction was slower and the tracklayers did not reach the valley-bottom town until mid-January 1907. Ballasting followed behind the track construction and took some time to complete, and scheduled passenger service between Oroville and Molson started on February 24, 1907.[11] Improvements in service continued with the much-appreciated introduction, in June 1907, of buffet-parlour car service as far west as Molson.

The steep descent into the Okanogan Valley took its toll. On April 18, 1907, a train loaded with rail ran away on the Molson Hill and rolled out of control down into Oroville, where it smashed into an engine from a track-laying train stopped in the yards. The crew of the parked engine was caught by surprise. Fireman Dan Evans was killed and four other railroaders were injured.[12]

As the track-laying progressed towards Oroville, work on clearing and grading the right-of-way had been advancing to the west of the valley. It led in a northwesterly direction following the Similkameen River to cross the border at Chopaka, west of Nighthawk, Washington, and continued into British Columbia to the communities of Cawston and Keremeos and eventually to the mining camps of Hedley and Princeton. Construction was not difficult although bridges over the Similkameen delayed the work. One tunnel not far west of Oroville was required. The border at Chopaka, B.C., was reached by the track-laying crew at 6:15 a.m. on April 10, 1907, and they proceeded on northward through the dry valley lands bordered by steep mountain ridges and unstable talus or scree slopes on the way to Keremeos.[13]

May brought warm weather to the mountains and high water to the Similkameen River, and three years' accumulation of driftwood and debris tumbled down the river as the surging water rose three feet (almost a metre) in just a few hours. The big Howe truss bridge at Armstrong (named after the nearby ranch and not to be confused with Armstrong in the north Okanagan) for which the piledriver crew had just finished the approaches, was damaged and so too was another major bridge. The damage took some time to repair but on July 10, 1907, the tracks reached Keremeos. Station buildings and telegraph lines were being constructed, and soon after a water tank, temporary coal dock and a four-track yard were built. A gravel pit near Keremeos supplied fill and ballast for the slowly advancing railway.[14]

The story of the construction of any railway is often like a play being presented on two stages. While the physical labour goes on in distant locations, management, financing, purchasing and politics go on that control the pace and direction of building. In 1907, there were several important changes in the management of the Great Northern and to its financial structure. James J. Hill resigned as president to become the chairman of the board. Louis Hill, capable, bright and more polished than his father, assumed the presidency. Nonetheless, James J. Hill was still very active in the long-term decisions and operations of his railroads. Also, the company

Grading a railway in the early 1900s was still hard, laborious work and required the recruitment of hundreds or thousands of men. Much of the work was done by hand and with horses, but increasingly steam shovels and other machinery became available to speed up the process and reduce the costs of labour. This photo shows grade workers, hired by one of the contractors, using teams of horses to construct the VV&E grade along the dry hillsides of the Similkameen Valley near Keremeos.

–Rev. Cameron, Joe Harris collection, Okanagan Archive Trust.

Oroville, just south of the international boundary, was a small agricultural community when the GN reached the Okanogan Valley from Molson. At right, GN train No. 255, with an E-6 Tenwheeler, is just pulling away from the depot, also shown above, in 1913. The railway extended north up the Similkameen River in the centre background, to Nighthawk, Washington, and on into British Columbia. Below, at left, two freights meet at Circle City on Molson Hill in 1907. The westbound in the foreground is headed by G-3 4-8-0 No. 725, while the train in the background is double-headed by a Class D 2-6-0 and an F-1 2-8-0. Westbound freights had to stop here to cool their wheels on the long descent. They also stopped at Nine Mile and Mount Hull. In the photo at far right, Engine 707, another G-3 4-8-0, is at Molson in 1907. These engines were limited to 425 tons on the Molson Hill. Molson was the highest station in the state at 3,703 feet (1129 m) elevation.

–Jacob S. Myers, three photos at right, Dave Wilkie Collection; above, Blair Photo, Brian Wilson Collection.

reorganized its many subsidiaries. By July 1, 1907, the GN had consolidated its investments in 14 affiliated lines in the United States and held the stock of the eight related British Columbia railways.[15]

Louis Hill, newly elected president of the Great Northern, arrived in Keremeos on July 16 on a tour of inspection and carried on by automobile to Hedley and Princeton to the surprise of many residents and miners alike. Hill drove the car himself, "on a road that required nerve on the part of the man at the helm. In him, *The Hedley Gazette* prophesied, "the Great Northern will continue to have, as it has always had, a virile head, for he is bound to prove a worthy scion of a worthy sire." Hill made few promises about the pace of construction to Hedley, pointing to a severe shortage of labour and the extent of Great Northern construction work west of the Rockies where, he noted, 8,000 men were employed.[16]

Hedley residents had a long wait for its railway. Grading progressed slowly from Keremeos along the Similkameen River towards Hedley and Princeton. By May 1908, as many as 200 men were working on the grade under the direction of George Cunningham, the superintendent of construction. J. Edward Kennedy, the son of the Great Northern's chief engineer, was chief engineer of construction on the VV&E. In August, camps were set up west of Hedley. Two large rock cuts near the camp at Allison, a few miles east of Princeton, were estimated to require seven months' work to complete.[17] Although the Great Northern was clearly committed to finishing this line, it was not rushing construction. By mid-May 1909, 80 percent of the grading had been completed between Hedley and Princeton.

The Great Northern's new branch line required several bridges, and during the summer of 1909, three piledrivers, two steam-powered and one, anachronistically, horse-driven, were employed on their construction. The steam piledrivers were at work on bridges over the Ashnola River and at Brushbottom while the horse-driven machine was employed at Hedley Creek where 1,000 feet (over 300 m) of pilings were needed. An interesting feature of the Howe truss bridges between Oroville and Princeton was the use of wooden sheeting, applied several years later, to the trusses.[18] This protective woodwork gave the bridges the appearance of being fully covered, but the tops were not roofed over. Three of this type, each requiring three spans, were used to cross the Similkameen River between Nighthawk and Hedley. The first, called the "Armstrong" bridge, was halfway between Cawston and the border; the second, Bridge No. 400 at Mile 163.8, was just west of Keremeos; and the third, Bridge No. 414, was located at Mile 170.7, midway between Keremeos and Hedley. Another set of three Howe truss bridges was required at Mile 124.5 just west of Oroville.

Track-laying west of Keremeos began in June 1909 and on August 13 reached Hedley, or "Smelter Flat," where the local residents had so long and eagerly awaited the arrival of the railway. By this time the Nickel Plate and Sunnyside holding was considered one of the "largest gold mines in Canada" and was operated under the name of the Hedley Gold Mining Company.[19] The large reduction works, owned by the Daly Reduction Company, became a landmark for the community for many decades to come.

By early October, steel-laying had reached Allison just a few miles east of Princeton, and by mid-month, residents of Princeton could hear the whistles of locomotives on the construction trains. On November 9, 1909, the tracks at last reached Princeton, and early in December, the line between there and Keremeos was inspected, permitting regular operations to begin. On December 17, the first inbound car of freight was delivered to Princeton over the long-awaited VV&E. Passenger service began on December 23, 1909, and regular mail and express service was introduced on January 1, 1910. Train No. 396, westbound, was scheduled to leave Oroville at 8:00 a.m., reach Keremeos at 10:30, stop briefly in Hedley one hour later, and finally arrive in Princeton at 1:00 p.m. Returning, Train No. 397 normally left at 2:30 p.m. and reached Keremeos at 4:20 p.m. and Oroville at 6:20 p.m.[20]

While construction was proceeding through the upper Similkameen and Tulameen valleys, VV&E

Customs and the Border...

"Grand Forks is the Custom port for the whole Boundary, with subports at Phoenix, Greenwood, Laurier, Midway, Sidley, Osoyoos and Keremeos, which report to R.R. Gilpin, the Collector of Customs."

The frequency of travel back and forth across the border was usually informal and easygoing in the pioneering days. As traffic became more commercial, formalities increased and added to the administrative work for shipments on the railways, particularly on the Great Northern with its many border crossings.

–*Henderson's British Columbia Gazetteer and Directory,* 1905.

On to Hedley and the Similkameen...

"...today the [Hedley] camp ranks as the most important in the whole Similkameen district: contains the largest gold mine in all Canada and gives every promise of still greater development when transportation is improved and mining facilities cheapened."

–Charles Camsell, *Geology and Ore Deposits of Hedley Mining District,* 1910.

Railway construction requires an enormous amount of fill and ballast. To satisfy the needs of the advancing VV&E through southern British Columbia, the engineers established a ballast pit near Keremeos and used a steam shovel to excavate the gravel. Such material was sometimes called "pit run" ballast, because it was given minimal treatment before being used on the railway. More expensive ballast was made from crushed rock or screened to remove fine material that disrupted drainage and consequently hastened tie deterioration, and it also encouraged the growth of weeds.

VV&E engineers also used steam shovels in grading and heavy construction along the line through the Similkameen Valley. These views show one of the shovels excavating gravel that is being loaded into a narrow gauge construction train. Materials were removed from cuttings along hillsides and then dumped nearby to fill depressions, in a technique that was called "cut and fill." These earthworks proved remarkably enduring and decades after the Great Northern tracks were removed many of these embankments remain in excellent condition.

–All photos on these pages by Rev. Cameron, Joe Harris Collection, Okanagan Archive Trust. Top right may be originally by Photo Supply Co., Spokane.

crews were also at work far to the west in the Fraser Valley, and improving connections to Vancouver. In 1908, VV&E construction included nearly 28 miles (45 km) of line from just east of Cloverdale to Huntingdon–Sumas on the B.C.–Washington border, just south of Abbotsford, providing connections with the Northern Pacific in Washington.[21] Within a few years, tracks were also extended eastward from Abbotsford to Kilgard and to within about a mile of the projected route of the Canadian Northern at Cannor, named for the Canadian Northern, on the south bank of the Fraser River. Improvements to the Great Northern's route between Vancouver and Seattle also came at this time. A new coastal route connecting Blaine, Washington, with Brownsville, across the Fraser from New Westminster, was opened in 1909. This was a high-priority project to replace a rather inefficient earlier connection through Surrey that was not well suited to high-speed services or handling significant traffic volumes.

Also in 1908, but in the southeastern corner of the province, the Great Northern's Crow's Nest Southern line into the Crowsnest coalfields was extended by 17.7 miles (28.5 km) from Fernie to Michel to access more high quality coal and coke for export to the United States

Keremeos at Last...

When the Great Northern finally arrived at Keremeos on July 10, 1907, it was a big event for the small farming, fruit growing and ranching community that was becoming established in the fertile dry valley of the Similkameen. The pioneers had much at stake, with the railway a critical link to the markets they needed. A display, featuring some very large melons, lined the station platform. Everyone dressed up for the event. The special and the reception are at left. Below, a businessman hurries from the coach not long after on what could be Louis Hill's inspection trip of July 16, 1907. Keremeos grew and the farms did well; the area still produces some of British Columbia's best fruit and produce.

The Great Northern at Hedley...

The GN came to Hedley in 1909 primarily to serve the lucrative Nickel Plate gold mine and the Hedley Reduction Works built in 1904. It was originally called the Daly Reduction Works, and dominated the dry, very steep and narrow valley of the Similkameen River between Keremeos and Princeton.

The reduction works processed the ore from the mine high above, reached by the tramway climbing up the steep mountainside behind. T. H. Rotherham photographed the arrival of the first passenger train, behind GN 2-6-0 463, at Hedley, upper right, on December 23, 1909, in what was undoubtedly a welcome Christmas present for the local population. The middle photo shows a train arriving at Hedley with the concentrator in the background, while in the lower photo a train from Oroville steams across the impressive 784-foot (239-m) trestle at Mile 178.8, just past the station, with the dam on Twenty Mile Creek in the foreground. Tracks were lifted west of the bridge to Princeton in 1940.

–FOUR PHOTOS, OKANAGAN ARCHIVE TRUST: WILSON COLLECTION, ABOVE; AT RIGHT, P. WRIGHT COLLECTION, COURTESY DOUG COX, TOP; WILSON COLLECTION, MIDDLE; HERB CLARK COLLECTION, COURTESY DOUG COX, BOTTOM.

The Nickel Plate Mine and the Hedley Reduction Works

The Nickel Plate Mine became one of the most famous large-scale lode gold workings in British Columbia. Its location, and in particular the imposing sight of the reduction works, overlooking the major highway through the southern Boundary country of the province leading to the Crowsnest Pass, made it a landmark for thousands of travellers in the post Second World War era.

The property was acquired by Marcus Daly and after his death in 1900 it was operated as part of his estate. Originally the company was known as the Yale Mining Company, but it was restructured as the Daly Reduction Company. In 1909, the Daly estate sold a substantial interest in the property and it became the Hedley Gold Mining Company.

Most of the claims were staked in 1898 and 1899 and the ore at Nickel Plate was first mined by glory holes, where large pockets of ore were removed. However, when they became too big, tunnels and shafts were driven and the ore was removed using stopes. For large parts of the workings, no timbering was needed because the rock was very hard and stable. At the nearby Sunnyside Mine, more care was required because in the spring the mine could be quite wet from seepage coming from the melting snow on the surface. When the mines were fully developed, ore was moved over 1.5 miles (2.4 km) of electric narrow gauge railway to a three-rail gravity tramway that lowered the ore cars, or skips, down to the reduction works in the valley. Grades on the tramway varied between 10 and 66.8 percent and, a 1910 study noted, it could handle about 150 tons of ore in a 10-hour shift.

The reduction works, completed in May 1904, were set against the base of Nickel Plate Mountain, which made it possible to move the ore through the works largely by gravity. After delivery from the tramway, the ore was passed over two grizzlies, coarse screens that removed oversize pieces of ore or rock, and then through two Farrell-type crushers. A conveyor then carried the ore to bins above eight batteries of stamps that broke up the chunks. Each of the 40 stamps weighed 1,050 pounds (475 kg) and dropped about 100 times a minute to crush the ore; the noise would have been deafening. The crushed ore passed to Homestake mortars, which further ground it down. From there the ore passed over copper amalgamating plates, where it was combined with mercury to form an amalgam. Then, while in suspension, it moved to Frue vanners that separated the ore according to weight into fine and coarse materials. The concentrates that were recovered, mostly arsenopyrites, were sacked and shipped by rail to Everett, Washington, for refining. The tailings were treated in a cyanide plant that consisted of a series of settling tanks where the heavier gold-bearing rock particles were collected. Residue was sluiced into the creek and into extensive disposal fields.[22]

Power for the mill was provided by damming Twenty Mile Creek upstream of the mill and building nearly three miles (4.8 km) of flume down to the mill and adjacent powerhouse. Compressed air was produced by a Rand compressor at the power plant. An air pipeline, 18,000 feet (nearly 5500 m) in length, delivered air, at 100 pounds per square inch pressure (689 kPa) to the mines. A hydroelectric plant provided electricity for the mines, the tramway and the townsite.

The mine proved to be profitable and long-lived. With its head offices far away in New York, the Hedley Gold Mining Company's annual report for the year 1911, for example, noted that with ore coming from the Nickel Plate, Sunnyside No. 4 and Silverplate workings, profits were $318,152.21 on 57,815 tons of ore milled, with dividends of 25 percent issued on the capital stock. In 1914 ore treated by either concentration or cyanidation reached 78,494 tons with some ores being valued at $20.00 a ton, and a 25 percent dividend continued. The last dividend was in 1919, however, but in the 1920s production continued. The Nickel Plate Mine's reserves were becoming depleted and after extensive drilling, the mine closed in 1930. The property was sold and then subsequently was acquired by the Kelowna Exploration Company based in New York, which found more ore. Production resumed in 1934, and by the mid-1940s production was estimated to have been worth at least $40 million. Mining continued at the Nickel Plate until 1955 when the mine closed. Other claims in the area also were developed, the most famous of these being the Mascot Fraction, which became very productive, and a mill was built to process its ore in the early 1930s. The Hedley Mascot Gold Mines Ltd. was very profitable and continued to produce high-grade ores into the post Second World War era, but closed in 1949. With the closure of the Nickel Plate Mine in 1955, the GN had little traffic left and abandoned its trackage to beyond Keremeos.[23]

The electric three-foot-gauge mine railway linked the surface and underground workings with the long tramway that conveyed the ore to the reduction plant. It was designed for low clearance in the mine workings. Two mine electrics and 40 ore cars were in use by 1917.
–BC Archives, 22377.

The Mascot Mine high above Hedley closed in 1949, but when demolition threatened it was saved by the provincial government with the support of heritage minister Bill Barlee. It reopened in the mid-1990s as a heritage attraction featuring a staircase leading to the precariously perched mine. It is operated by the Upper Similkameen First Nation with spectacular tours available from the Snaza'ist Centre each summer.
–Hugh Fraser.

Andrew McCulloch and James J. Warren, seated, with Kettle Valley staff and officials at Penticton, pushed the line across the southern British Columbia mountains.
–Penticton Museum.

Uncle Sam Opens the Colville Reservation....

"350,000 acres of desirable agricultural lands of the south half of the Colville Indian Reservation–located in the Columbia River and Okanogan valleys of north-central Washington–are to be disposed of to settlers. 'Go Great Northern' and have your choice of five registration points including Omak, the only registration point actually on the reservation, and reached only by the Great Northern."
–Great Northern Timetable, June 1916, Robert Turner Collection.

Princeton. The first major obstacle was a long ridge just beyond Princeton that had to be tunnelled to permit the tracks to follow the Tulameen River to the northwest. By the end of May 1910, as crews worked at both ends, about one third of the 1,062-foot (324-m) bore had been finished. However, construction went slowly that summer and in November the GN crews stopped work for the winter with grading not yet as far as Coalmont.

In June 1911, grading crews reached Tulameen and in late July track-laying started through the Princeton tunnel, but it would take until the autumn of 1911 before all of the bridges, which included five crossings of the Tulameen, were in place. The tracklayers, working for contractors A. Guthrie & Co., reached Coalmont, 13 miles (21 km) beyond Princeton, on November 10. The trackage was opened for regular service on May 1, 1912.[28]

Slowly, but surely, the GN was closing the gap between the eastern sections of the VV&E and Vancouver. Meanwhile, the Canadian Pacific finally was pushing ahead with construction of the Kettle Valley Railway, working south from Merritt to Brookmere to the southeast, and from Midway westward to Rock Creek and over the rugged mountains and Myra Canyon, which fronted the eastern side of the Okanagan Valley.

While construction crews laboured in the mountains of southern British Columbia, events perhaps more important to the fate of the railways were unfolding on a national level, as described earlier. Following the defeat of Wilfrid Laurier's government on September 21, 1911, any immediate hopes for reciprocity–free trade between Canada and the United States collapsed. For Hill, it ended any hope of free trade developing between the two countries and undoubtedly brought an end to his visions for the VV&E and much of the underlying potential and justification for the trackage west of the Boundary mining districts. As Ralph Hidy and his co-authors in their study of the Great Northern comment, by this time, "Hill apparently had abandoned this dream [of a route across western Canada] and had decided to settle for alliances with CPR's Canadian competitors." Despite the political defeat of free trade, the Great Northern continued on with construction to complete the VV&E, while rival CPR, undoubtedly encouraged by the same political change, pushed across the southern interior with the Kettle Valley Railway. But from that point on, the competitive fire was gone and a more cooperative approach was taken to completing the very expensive railway through the coastal mountain ranges.

By the autumn of 1912, a contract was let by the Great Northern to Guthrie, McDougall & Co. for grading from Coalmont towards Coquihalla Summit. However, by this time an agreement had been reached between the Great Northern and the Kettle Valley to share trackage between Princeton and Hope, saving both companies heavy, duplicate and very expensive construction. Under the agreement, which took some time to formalize in all its many details, the GN was to complete trackage as far as Otter Summit, soon renamed Brookmere, 38 miles (60 km) northwest of Princeton at milepost 240.7. From that point west, the joint trackage agreement utilized KVR trackage to Brodie Junction and from there through the Coquihalla Pass to Hope. At Brodie, the Kettle Valley, which was then being built upstream along the Coldwater River from Merritt, was joined.[29]

The agreement provided for the payment to the Great Northern of a yearly rental fee of one twelfth of 2.5 percent of the total costs of construction and a similar annual fee for additional improvements to the property. The GN was responsible for the construction and upkeep of stations and other structures. Maintenance and related expenses including snow removal were to be shared on a "pro rata proportion of the cost incurred..." based on the number of miles run each month by engines and cars of each company.[30] A similar agreement provided for payments, on an annual basis, to the CPR by the Great Northern for operating rights over the Coquihalla Pass. On an annual basis, the GN received $60,000 from the CPR for use of the 38 miles (61 km) of trackage from Princeton to Brookmere while the CPR was paid $150,000 a year for use of the more expensive 56.9 miles (91.6 km) of the Kettle Valley Railway through the Coquihalla.

The VV&E route proceeded almost due north

from Tulameen along the eastern shore of Otter Lake to the head of the Otter Creek drainage. From there it crossed over to follow the Spearing Creek drainage to Brookmere. Grades on this section were moderate, reaching a maximum ascending gradient of 1.2 percent to Brookmere. With the joint trackage agreements in place, GN construction ended at Brookmere.

Brookmere, deep in the mountains, was to be an important terminal on the railways where crews changed and helper locomotives, required on the grades over Coquihalla Pass, were based. This remote station was reached by KVR crews in 1911, but it had remained the end of track pending the decision on the Kettle Valley routes to the east and west. GN track-laying was completed to Brookmere in mid-October 1914. However, the Great Northern had no immediate reason to operate past Tulameen and the CPR did not complete its trackage between Penticton and Princeton, over the rolling mountainous highlands, until April, the following spring. Finally, on May 28, 1915, the trackage to Brookmere was approved for operation. Three days later, amid celebration and fanfare, passenger service began on the Kettle Valley Railway between Merritt and Midway.[31]

For the Great Northern, the completion of construction to Brookmere had little immediate impact because there was no through traffic or connections westward until work was finished through the Coquihalla. What little local traffic there was beyond Princeton, the GN mostly turned over to the Kettle Valley after May 31, 1915, or GN cars were moved under a contract in effect between July 1, 1916 and Feburary 15, 1931.[32] However, the GN did operate on a seasonal basis as far as Tulameen during the winter when ice was harvested and carried south to Washington.

The arrival of the Great Northern at Hedley, Prineton and Coalmont provided these areas with a long-desired modern transportation route. The Nickel Plate Mine at Hedley was certainly one of the prizes for the VV&E, and it continued to be a major source of traffic for the GN. In 1914, the mill processed 78,494 tons of ore valued, after recovery of the gold, at $797,340.76. Because

periodic water shortages continued to hamper production, in 1913 the mining company began work on the construction of a hydroelectric plant on the Tulameen River to provide power for the mine and the town. This plant was opened on January 2, 1915. At Coalmont, development work was underway on the coal seams situated near the Great Northern. Two years earlier, in 1913, this property, controlled by the Columbia Coal & Coke Company, had been acquired by Arthur McEvoy of Vancouver and small quantities of coal were shipped out over the Great Northern, but the property never developed into a major producer. After the completion of the railway through the Coquihalla, described later, Coalmont traffic was moved to Princeton by the Canadian Pacific, and at times the volumes could be significant, testing the capacity of GN's tri-weekly service, sometimes requiring the railway to operate extra freights to Princeton from Oroville.[33]

Meanwhile, world events were overtaking local and national considerations in determining the prosperity of the VV&E, the Canadian Pacific's branch lines and the mining districts. In August 1914, the First World War broke out in Europe, Canada was soon fully engaged, and the economics and politics of railway developments and the lives of thousands of people in the mining districts changed forever.

OROVILLE AND PRINCETON

397 Mixed Tues. and Fri.	Mls	Table 74 Pacific Time	396 Mixed Tues. and Fri.
1 30	0	Lv **Oroville** *86*....Ar	9 30
2 14	11	" Nighthawk....Lv	8 45
2 55	21	" Chopaka, Wash. "	8 06
		International Boundary	
3 33	31	" Similkameen, B.C. "	7 28
f 3 47	34	" Cawston...... "	f 7 15
4 00	38	" Keremeos..... "	7 00
— —	45	" Ashnola....... "	— —
— —	51	" Bradshaw..... "	— —
5 30	56	" Hedley....... "	5 35
....	60	" Cory......... "	....
....	68	" Bromley...... "	....
....	72	" Norman...... "	....
....	77	" Allison........ "	....
....	80	Ar **Princeton, B.C.** Lv	....

A twice-weekly mixed train was all the scheduled service provided between Oroville and Hedley by the mid-1920s. Passenger service to Princeton had already been suspended. It took four hours, officially at least, to travel the 56 miles (90 km) between Oroville and Hedley. But mixed services on rural branch lines were notoriously slow and late, because freight cars had to be switched all along the runs. This excerpt is from the Great Northern's October 1, 1924, timetable.
–Robert Turner Collection.

The town of Coalmont along the VV&E, like so many others, had high hopes of mining prosperity but the coal deposits proved disappointing, and use of coal declined generally and as the GN converted its engines to burn oil. Coal from Blakeburn, where mines operated from 1918 to 1940, was moved to Coalmont over a 3-mile (5-km) aerial tramway and shipped over the Kettle Valley. Some of this coal also went south over the GN from Princeton.
–Priest photo, Wilson Collection, Okanagan Archives Trust.

Looking back over several generations it is difficult to imagine how small the communities along the new railways actually were in the late 1890s and early 1900s. A town like Keremeos, which clamoured for the railway, had only a few hundred people when the GN arrived in 1907. Not many people could afford to travel, although the railway meant a great deal for its connections to the outside world for mail and supplies and of course as a way to reach markets for crops. Isolation was at an end when the railway arrived. This photo is from about 1920.

For a few years a first-class service operated from Spokane and Marcus through to Oroville via Grand Forks and Molson. At right, eastbound train No. 255 pauses at Curlew behind GN's E-6 class 4-6-0 925. The first car is a Railway Post Office or RPO car. Mail contracts were a key source of revenue in supporting passenger train operations.

–Robert Turner Collection, above; G. E. Thomas photo, Warren Wing Collection, at right.

SPOKANE, GRAND FORKS AND PHOENIX.

	No. 256 Ex. Su.	No. 390 Mixed Ex. Sun.	Miles	Table No. 55 Pacific Time	No. 389 Mixed Ex. Sun.	No. 255 Ex. Su.	
	8 55 AM		0	Lv. 6 **Spokane** 58. Ar		**5 30** PM	
	1 15 PM		101	**Marcus** 58. 60		**1 05** PM	
	3 15 PM		143	Ar. **Grand Forks**. Lv		11 10 AM	
		3 30 PM	143	Lv **Grand Forks** 58 Ar	10 20 AM		
		— —	144	Weston	— —		
		x **4 02** PM	150	Spencer	x 9 58 AM		
		x **4 27** PM	157	Hale	x 9 33 AM		
		x **4 32** PM	158	Denoro	x 9 27 AM		
		x **4 40** PM	160	Glenside	x 9 22 AM		
			166	Jordan			
		5 10 PM	167	Ar. . . **Phoenix** . . Lv	9 00 AM		

The Grand Forks-to-Phoenix service was a daily-except-Sunday mixed train in the summer of 1916, scheduled for an hour and a half for the downhill run over the 24-mile (39-km) branch line. Climbing the grade took an extra 10 minutes, if all went well.

READ DOWN **SPOKANE, MARCUS, GRAND FORKS AND OROVILLE.** READ UP

No. 262 Sunday Only	No. 258 Daily Ex. Su.	No. 256 Daily Ex. Su.	Miles	Table No. 54 Pacific Time	No. 255 Daily Ex. Su.	No. 257 Daily Ex. Su.	No. 261 Sunday Only
8 55 AM	**3 55** PM	8 55 AM	0	Lv **Spokane 6, 65** Ar	**5 30** PM	11 35 AM	**8 05** PM
9 10 AM	— —	9 10 AM	5	Hillyard	**5 15** PM	11 20 AM	**7 45** PM
9 30 AM	**4 33** PM	9 30 AM	14	Dean	**4 55** PM	11 00 AM	**7 25** PM
x 9 40 AM	x **4 42** PM	— —	18	Wayside	— —	x10 50 AM	**7 12** PM
x 9 47 AM	x **4 51** PM	— —	20	Darts	— —	x10 43 AM	x **7 03** PM
x 9 53 AM	x **4 59** PM	— —	23	Denison	— —	x10 38 AM	x **6 58** PM
10 02 AM	**5 10** PM	10 02 AM	27	**Deer Park**	**4 22** PM	10 30 AM	**6 50** PM
10 15 AM	**5 23** PM	10 15 AM	32	Clayton	**4 06** PM	10 15 AM	**6 36** PM
10 35 AM	**5 38** PM	10 35 AM	38	Loon Lake	**3 50** PM	9 57 AM	**6 20** PM
x— —	x— —	— —	43	Holland Horr Spur	— —	x— —	x— —
10 55 AM	**5 58** PM	10 55 AM	47	Springdale	**3 27** PM	9 32 AM	**5 50** PM
x11 05 AM	x **6 09** PM	— —	51	Grays	— —	x 9 20 AM	x **5 36** PM
x— —	x— —	— —	54	Kulzers	— —	x— —	x— —
11 14 AM	**6 22** PM	11 14 AM	56	Valley	**2 54** PM	9 09 AM	**5 21** PM
11 29 AM	**6 39** PM	11 29 AM	64	**Chewelah**	**2 39** PM	8 50 AM	**5 05** PM
x— —	x— —	x— —	70	Blue Creek	— —	x— —	x— —
11 49 AM	**7 00** PM	11 51 AM	73	Addy	**2 19** PM	8 30 AM	**4 43** PM
x**12 04** PM	x **7 15** PM	x— —	81	Arden	x— —	x 8 13 AM	x **4 25** PM
x**12 10** PM	x **7 24** PM	— —	84	Kiel	— —	x 8 06 AM	x **4 18** PM
12 17 PM	**7 35** PM	**12 20** PM	87	**Colville**	**1 50** PM	7 58 AM	**4 10** PM
x— —	x— —	x— —	94	Mission	x— —	x— —	x— —
12 40 PM	**7 55** PM	**12 40** PM	96	Meyers Falls	**1 28** PM	7 39 AM	**3 50** PM
12 55 PM	**8 10** PM	**12 55** PM	101	Ar. . . . **Marcus 55** . . Lv	**1 05** PM	7 20 AM	**3 30** PM
		1 15 PM	101	Lv. . . **Marcus** . . . . Ar	**12 50** PM		
		x **1 30** PM	106	Boyds	x**12 38** PM		
		x **1 42** PM	111	Barstow	x**12 27** PM		
		x **1 52** PM	116	Dulwich	x**12 16** PM		
		1 56 PM	118	Orient	**12 11** PM		
		2 35 PM	128	Laurier	11 46 AM		
		x **2 48** PM	134	Rideau	x11 30 AM		
		3 20 PM	143	**Grand Forks** 55	11 10 AM		
		3 40 PM	146	Danville, Wash	10 45 AM		
		x **3 50** PM	150	Hurlburt	x10 32 AM		
		4 05 PM	156	**Curlew**	10 20 AM		
		x **4 19** PM	162	Paxson	x10 02 AM		
		x **4 28** PM	166	Torodo	x 9 55 AM		
		4 40 PM	171	Ferry	9 45 AM		
		4 50 PM	171	Midway	9 38 AM		
		x **5 13** PM	180	Bergen	x 9 23 AM		
		5 29 PM	185	Myncaster	9 10 AM		
		x **5 40** PM	190	Syackan	x 9 00 AM		
		6 00 PM	197	Bridesville	8 45 AM		
No. 396 Mixed Tu., Th., Sat.		**6 15** PM	202	Molson	8 30 AM		No. 397 Mixed Tu., Th., Sat.
		x **6 35** PM	209	Nine Mile	x 7 55 AM		
		6 55 PM	213	Circle	x 7 35 AM		
		x **7 16** PM	220	Mount Hull	x 7 00 AM		
		7 40 PM	227	Ar. . . . **Oroville** Lv	6 30 AM		
7 00 AM			227	Lv. . . . **Oroville** . . . . Ar			**6 20** PM
x— —			229	Rich Bar			x— —
7 45 AM			238	Nighthawk			**5 42** PM
x— —			240	Ruby Mine Spur			x— —
8 30 AM			248	Chopaka			**5 06** PM
x 9 00 AM			258	Similkameen			x **4 21** PM
9 30 AM			265	Ar. . . **Keremeos** . . . Lv			**3 51** PM
x 9 55 AM			272	Ashnola			x **3 21** PM
x10 15 AM			278	Bradshaw			x **3 01** PM
10 30 AM			283	Hedley			**2 46** PM
x10 45 AM			288	Cory			x **2 21** PM
x11 05 AM			295	Bromley			x **2 06** PM
x11 20 AM			299	Norman			x **1 52** PM
x11 30 AM			304	Allison			x **1 37** PM
11 45 AM			307	Princeton			**1 30** PM

Equipment.—Coaches and Smoking Cars between Spokane and Oroville, Buffet-Parlor Observation Car between Spokane and Republic, on Nos. 255 and 256.

SPOKANE, CURLEW AND REPUBLIC.

No. 392 Mixed Ex. Su.	No. 256 Daily Ex. Su.	No. 394 Mixed Ex. Su.	Miles	Table No. 56 Pacific Time	No. 391 Mixed Ex. Su.	No. 255 Daily Ex. Su.	No. 393 Mixed Ex. Su.
	8 55 AM		0	Lv. 54 **Spokane** 65. Ar		**5 30** PM	
	1 15 PM		101	**Marcus**		**1 05** PM	
	4 05 PM		156	Ar. . . . **Curlew** Lv		10 20 AM	
10 25 AM		**4 10** PM	156	Lv. . . . **Curlew** . . . . Ar	10 10 AM		**3 50** PM
x10 42 AM		x **4 24** PM	162	Malo	x 9 53 AM		x **3 20** PM
x— —		x— —		Karamin	x— —		x— —
x11 10 AM		x **4 41** PM	169	Poliard	x 9 35 AM		x **2 55** PM
x11 25 AM		x **4 52** PM	173	Torboy	x 9 25 AM		x **2 45** PM
11 40 AM		**5 10** PM	178	Ar. . . **Republic** . . . Lv	9 10 AM		**2 30** PM

Great Northern Railway timetables for summer 1916. The timetable noted that Great Northern eating house service was offered at Marcus.

–Robert Turner Collection.

Marcus, located at the junction of the GN's former Spokane Falls & Northern and its line to Republic and Grand Forks, was a busy town. In the lower photo, the big bridge crossed the Columbia on the line to Grand Forks, Curlew, Republic and Oroville. Three trains are at the station: one is from Oroville eastbound, one northbound to Nelson and another in from Spokane. At left passengers change trains across the Marcus depot platform.
–Nancy Denny Collection, below; Warren Wing Collection at left.

The GN's Third Main Line: Illusion and Reality

Promises Were Easy...

"'Would it be indiscreet, Mr. Hill, to ask how you are getting along with the Hope mountain proposition?'

'Not at all, not at all,' was the emphatic and pleasant answer. 'We shall get through to the coast all right. Difficulties of course have to be faced, but we shall find a good route. As you know we do not want to make three trains do what one should do.'"

—*The Grand Forks Gazette*, September 30, 1905.

Only One Great Northern Train...

"Louis W. Hill, president of the Great Northern Railway, came over the Kettle Valley Line from Vancouver and spent the night in his private car. Thursday his special train of four coaches proceeded to Spokane over the Great Northern."

—*The Similkameen Star*, Princeton, September 29, 1916.

Abandonments Were Difficult...

"We have a lot of lines in Canada which I think we all agree we would be glad to cease operating, it is at the present time impossible to do other than continue the operation of the Abbotsford–Cannor Line because our Canadian charters are based on its through operation, and if it would be worked out at all, it would be a long process."

—Great Northern President Ralph Budd to William P. Kenney, October 18, 1919 (Great Northern President's files, as quoted by David C. Jones, 1980, p. 148).

Too Optimistic, by About Six Years...

"Jim Hill and the VV&E will be in Merritt within 12 months according to the News. It is understood on the best of authority that the railway magnate has made up his mind to get in touch with the coal fields of Nicola and run a spur from the junctional point at the head of the Coldwater.

"Over 2,000 men are now at work in the vicinity of Tulameen pushing the construction of the road across the Hope Mountains. They are making splendid progress and the VV&E will be in operation into Vancouver before many months have passed."

–*The Phoenix Pioneer,* July 9, 1910.

The Great Northern had taken a slow pace through the Fraser Valley because there was little traffic potential in local services, and any through service had to await the completion of trackage from the east through the mountains. By March 1911, reports quoting Chief Engineer Kennedy stated that the GN intended to use the trackage of the as yet incomplete Canadian Northern Pacific in the Fraser Valley beyond the Sumas River, between a junction to be named Cannor and Hope. This removed the immediacy of constructing the connecting trackage for a few years and the need to build to the community of Chilliwack. By mid-July 1912, only 4.52 miles (7.25 km) of track had been laid east of Abbotsford and the next year another 13 miles (21 km) was added, much of it on trestlework along the north shore of Sumas Lake. It took until September 15, 1916, for GN's connection to be made with the Canadian Northern at Cannor, 37 miles (60 km) west of Hope along the south bank of the Fraser. The GN agreement with the Canadian Northern was said to cost $50,000 a year.[1]

The CNP, which later became part of Canadian National Railways, was developed as a new transcontinental system by William Mackenzie and Donald Mann, who had also been early promoters of the VV&E. The Canadian Northern generally ran north of the CPR across the Canadian Prairies before crossing the Rockies through Yellowhead Pass west of Jasper. In British Columbia, the line was constructed under a provincial charter and took on the name Canadian Northern Pacific Railway. West of the Yellowhead Pass and the nearby landmark of Mount Robson, it turned south to follow the North Thompson River to Kamloops, then followed the Thompson and Fraser rivers south, usually being located on the opposite bank from the CPR. This brought the CNP to Hope (and its connection with the VV&E) and from there it followed the south bank of the Fraser to a point east of New Westminster named Port Mann where terminal facilities were built. From Port Mann, the line continued west to cross the river over the New Westminster bridge.

The Great Northern's entrance to Vancouver may have been expensive, but in the long term it proved a good investment and provided an important north-south rail corridor between Vancouver and Seattle. Opposite, a long GN freight, with 4-6-2 1723, leaves Vancouver for Seattle on May 16, 1940, through the filled lands at the eastern end of False Creek from the GN's station and yards. GN's line to Seattle followed the coastline nearly the whole way and permitted locomotives to handle long trains.
–Norman Gidney.

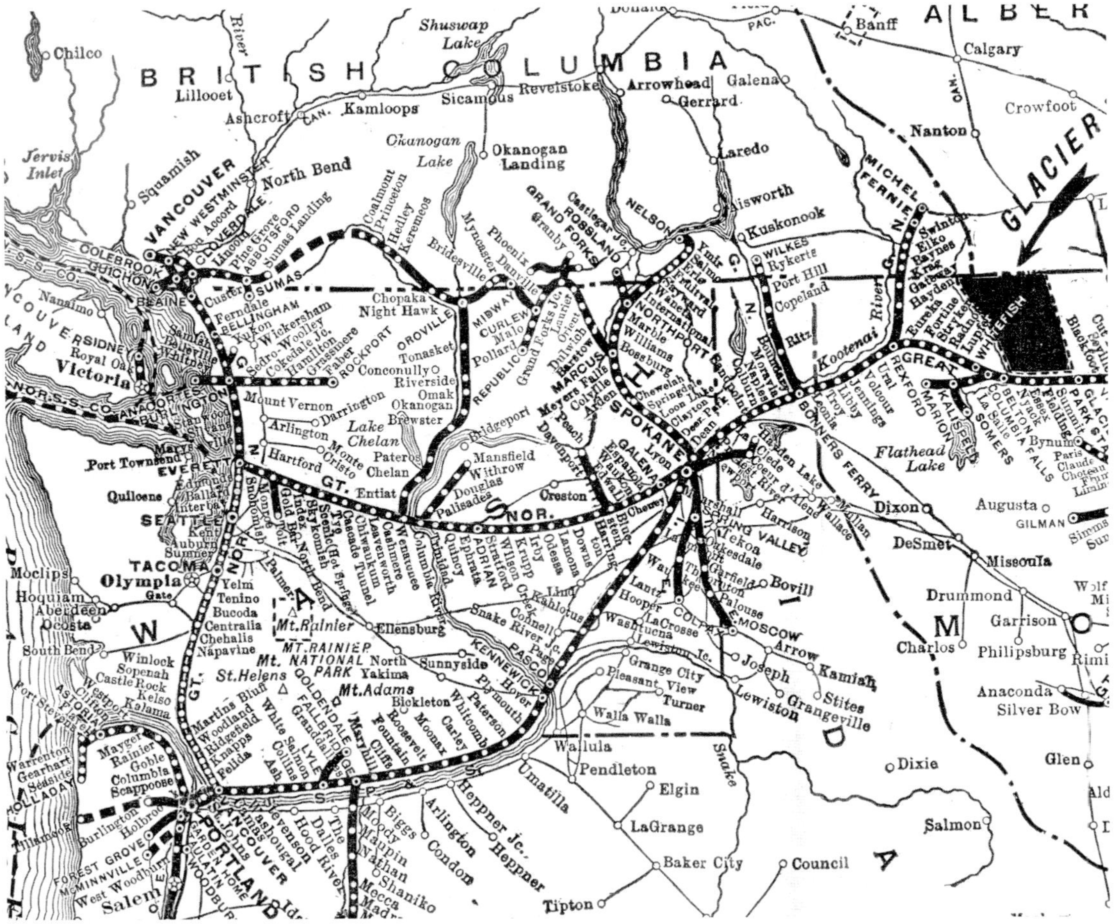

This map, from the Great Northern's public timetable in 1916, shows the railway's three routes to the coast: over the main line through the Cascades to Everett and Seattle; to the south, via the Spokane, Portland & Seattle along the Columbia to Portland; and the northern line over the serpentine VV&E and Washington & Great Northern from Spokane, still incomplete, through the mountains of British Columbia, to Vancouver.
–Robert Turner Collection.

followed a south-to-north, 137-mile-long (220-km) route connecting the major agricultural centres along the river including Pateros, Omak and Tonasket. The expansion of trackage in the Okanogan produced a twofold advantage: it provided an effective bypass for the Molson Hill and it expanded GN influence along the highly productive irrigation lands adjacent to the Okanogan and Columbia rivers. This railroad expansion reflected the growing importance of the area and James J. Hill's interest and investments in the region. Hill had actively promoted irrigation in the Wenatchee region and northward towards the B.C. border. He and other backers had purchased bonds in a number of key development companies in the area, and by 1905, some 30,000 acres (12 140 ha) of orchard lands were in production. The Great Northern also supported the West Okanogan Irrigation district and by 1916 another 7,000 acres (2830 ha) were being irrigated. The net result was that Washington's Okanogan–Wenatchee region, like British Columbia's Okanagan Valley to the north, was becoming a key fruit growing region, so much so that by 1917 it was producing 20 percent of the apples grown commercially in the United States.[8] At the same time, on the B.C. side of the border, orchards and farms were coming into production between Oroville and Keremeos, adding to the importance of agricultural products as a major source of traffic on the branch line.

This alternative route effectively eliminated any need for direct service between Vancouver and Spokane over the VV&E with its steep grades and winding mountain trackage, and took away the utility of some sections, particularly the Midway-to-Oroville trackage, as a branch line. Moreover, as Ralph Hidy and his colleagues concluded, "throughout 1916 and 1917, GN officials concentrated on meeting increased demands rather than generating new traffic." Critically, on April 6, 1917, the United States declared war on Germany and its allies, and the American railroads found themselves in a new and very troubled and demanding era. They faced escalating requirements and demands for war materials and manufactured goods, many ordered by the Allied powers in Europe, shortages of freight cars, poor equipment utilization, many bottlenecks, and an overall poorly functioning integration of the many railroad companies. To quote Hidy, "the GN shared in the rail carriers' ultimate failure to meet wartime demands [although by] voluntary cooperation, they had reduced passenger service, pooled freight equipment, eliminated much congestion at eastern terminals and minimized car shortages." Adding to the difficulties were an unwieldy pre-war regulatory structure, financing constraints and rate restrictions.[9]

This clearly was not a propitious time for an American railroad to be expanding services in Canada, when every effort was required to acquire new equipment and provide for services at home. In fact passenger

services in many areas were reduced and by the end of 1917 the situation across the United States had deteriorated further and reached a crisis. On January 1, 1918, President Woodrow Wilson announced that the federal government would take over control of America's railroads under what became the United States Railroad Administration (USRA). Soon after, William P. Kenney, a lifelong railroader and vice-president of traffic with the Great Northern, became federal manager (and briefly president) for the railroad after Louis Hill refused to work for the government. A few months later Hill resumed the presidency, and Ralph Budd assumed the role of chief executive officer, but the USRA controlled priorities and operations. Kenney, as manager and following USRA priorities, "practically stopped immigration and agricultural development work, discontinued solicitation of traffic, closed off-line ticket offices and did no promotional advertising."[10]

Meanwhile, before conditions brought severe restrictions on expansion, the Great Northern began developing extensive new terminal facilities in Vancouver, and, of course, efficient yards were a key to improving equipment utilization. Besides new freight and passenger yards, the major feature was an elegant and ornate new station built on reclaimed lands at the eastern end of False Creek next to what was to become the Canadian Northern's station and passenger yards. The new GN station, completed late in 1917, was designed by Frederick L. Townley of Vancouver, the architect who also designed Vancouver's city hall. It was an impressive structure, matching the railway's aspirations for its presence in the city. The main building, imposing and spacious at 235 x 60 feet (72 x 18 m), had terminal facilities on the main floor and offices above with an adjoining wing measuring 130 x 42 feet (40 x 13 m) for baggage, mail and express offices and rooms.[11] But this striking structure was really intended to serve the frequent passenger trains operating between Vancouver and Seattle, and any thoughts of it being the terminus for express passenger trains running via the southern interior of British Columbia to Spokane had long faded away like a distant and fleeting dream.

USRA management of the railroads continued until March 1, 1920, nearly 16 months after the armistice of

Great Northern's beautiful Vancouver station, next to Canadian Northern's, was a landmark of the city after its opening in 1917. The scene above from 1947 shows the station with a banner welcoming visitors to tour GN's new Seattle-to-Chicago streamlined *Empire Builder,* cars of which were on display. Eight years later Canadian Pacific unveiled *The Canadian* for its transcontinental service. Great Northern was an early advocate of colourful streamlined passenger trains and soon inaugurated the *Streamlined International* service between Vancouver and Seattle. Above left, GN's 1458 arrives in Seattle on the *International Limited* from Vancouver on July 26, 1946.
–Above, Ronald Nixon, No. 12694; above left, Albert Farrow.

Leisurely Mixed Trains...

SUMAS, CLOVERDALE AND GUICHON

Read Down — *Read Up*

383 Mixed Ex. Su.	Mls.	Table 86 Pacific Time	384 Mixed Ex. Su.
10 15	0	Lv **Sumas** .... Ar	9 15
— —	1	" **Huntingdon** .. Lv	9 14
10 30	4	" **Abbotsford** *83* "	9 00
10 45	8	" Sarel .... "	8 35
11 05	13	" Aldergrove .... "	8 20
11 40	17	" Otter .... "	7 55
12 05	22	" Lincoln .... "	7 25
12 55	29	" **Cloverdale** .... "	6 55
f **1 15**	33	" Alluvia .... "	6 40
f **1 25**	35	" Southport ..... "	f 6 35
1 35	36	Ar **Colebrook** *88* .. "	6 30
3 56	36	Lv **Colebrook** .... Ar	5 55
f **4 25**	43	" Inverholm Lv	f 5 15
f **4 45**	45	" Ladner .... "	f 5 05
5 00	46	Ar **Guichon** Lv	5 00

ABBOTSFORD AND KILGARD

398 Mixed Tues. & Fri.	Mls	Table 83 Pacific Time	399 Mixed Tues. & Fri.
9 00	0	Lv **Abbotsford** *86*. Ar	9 40
9 20	5	Ar Kilgard Lv	9 20

The Great Northern's Fraser Valley timetables, October 1, 1924.

–Robert D. Turner Collection.

Meanwhile, Canadian Pacific Continued to Improve Services...

"Café-dining car *Opinaka* now on the Nelson-Vancouver run has an observation platform, 16 chairs in the parlour compartment, latest electric light system and ventilating fans in each compartment. By the time summer schedule goes into effect six of these cars, one on each train will be in service. All sleeper passengers will have access to this car."

–*The Grand Forks Gazette,* May 19, 1917.

November 11, 1918. During this time Ralph Budd was elected president of the Great Northern, a role for which he had been groomed by the late James J. Hill. After the end of the USRA era he took full control of the railway. Budd proved to be an astute, effective and efficient executive for the railroad in the demanding years ahead that included a severe postwar recession in the early 1920s, and he continued with GN until 1931. Louis Hill remained on the railway's board until his death in 1948.

The Third Main Line, as it has sometimes been called, was to survive for a time, but just as a linked collection of branch lines useful only as long as there was sufficient local traffic to support them. The new Wenatchee-to-Oroville connection held one other key to the future of the Third Main Line: by making the trackage from Oroville to Tulameen effectively an extension of the Okanogan branch, it was possible to ship blocks of ice cut from Otter Lake, at Tulameen, directly to the fruit growing regions. The need of ice for storage and in refrigerator cars grew dramatically as thousands of cars were used to move central Washington's perishable produce to markets all over the continent. The branch line that was built to serve the mines of the Hedley–Princeton district was, for the next decade, intimately connected with fruit production in Washington. Additional information about the ice shipments is included in a later chapter dealing with the GN's railway in the Similkameen Valley.[12]

The GN's Fraser Valley lines functioned as local branch lines for some years but never became prosperous or productive. A mixed train service ran through the valley to Hope until about 1919, but in 1920 the junction with the Canadian Northern at Cannor was removed, although apparently without formal permission. By then the Canadian Northern had gone into bankruptcy and been incorporated into the government-owned Canadian National Railways. Most of the remaining Fraser Valley trackage survived until the Great Depression, when the rural branch lines were abandoned. High maintenance costs of the many long trestles through the westlands contributed to the demise of some sections. The traffic did not justify the high costs of reconstruction.[13]

These meandering lines in the Fraser Valley are in many ways another story, one of mixed trains handling local traffic on a very relaxed schedule that could not compete with the expanding road systems and increasing efficiency of automobiles. They also faced competition from British Columbia Electric's extensive, fast and efficient freight and passenger electric interurban railway system, and from Canadian Pacific and Canadian National for freight traffic and to an extent for the rural passenger business in the valley. Fortunately, the GN's north–south main line between Vancouver and Seattle prospered, but that too is another story, and one tied to the expansion of the urban and industrial development that spread around Vancouver and the major cities to the south along Puget Sound.

The abandonment of the notion of a continuous through route was really much more in keeping with James J. Hill's pragmatic philosophy of railway construction. The GN investments in its main line were long-term, energetic and thoughtful, and in all likelihood, if one were able to look behind the rhetoric and competitive hoopla and the ebullient, dynamic personality of James J. Hill, the so-called Third Main Line probably never really was a serious concept as much as it was an integration of branch lines fostering traffic for the Great Northern main line from resource development areas. And of course, it was a valuable piece in the chess game of competitive interests played out with the CPR and its importance soon passed: sometimes a bluff, sometimes an element of strategy, sometimes a win, sometimes a loss until the game was over. However, there was certainly no denying that the VV&E enterprise had been expensive and not very fruitful at least as an overall notion. A November 30, 1914, account put the total cost of the VV&E and related endeavours at $22,566,240.83. Of that about $7,700,000 was charged to the New Westminster-to-Vancouver line (built originally as the Vancouver, Westminster & Yukon) which included the massive excavation known as the Grandview Cut, and to Vancouver's extensive new terminals, a dock on Bur-

The rural charm of Great Northern's Fraser Valley branch lines was reflected in the elderly locomotives, the mixed trains, the aging wooden combine (coach and baggage car), and the small depots at the farming communities. Above left, GN's 2-6-0 No. 306, clean and polished, is serviced at Vancouver between duties in the Fraser Valley on the Colebrook Local in the late 1920s. The extra water tender extended the range of the engine. At right, the mixed train, this day without any freight cars, pauses at Aldergrove, west of Abbotsford, behind 2-8-0 No. 572. Engine 453 was another Fraser Valley veteran at Vancouver in the 1920s. The New Westminster Bridge over the Fraser gave GN, CN and B.C. Electric access to New Westminster and Vancouver.

–Engines, Ken Merilees; Above, H. T. Hayashi, Pacific Coast Division, Canadian Railroad Historical Assoc.; bridge, Robert Turner Collection.

rard Inlet and related trackage. There were additional charges, included in the total, of over $1,300,000 for improved connections (technically part of the Victoria Terminal Railway & Ferry Company) to Blaine, Washington, and between Cloverdale in the Fraser Valley and Port Guichon (near the present-day city of Richmond, B.C.). The overall account included construction costs between Midway and Molson, but not between Molson and Chopaka (part of the Washington & Great Northern) and this section came to $2,741,744.[14] While the total represented a substantial investment, some of the properties were indeed valuable and proved a long-term benefit to the Great Northern. This could certainly be said of the Vancouver terminals and the trackage between New Westminster and Vancouver from which the GN received substantial payments from the Canadian Northern and later the Canadian National as part of their joint trackage agreements for access to Vancouver.

Two years after the GN's route to the coast was completed between the mining districts and Vancouver, the First World War ended. Soon after, copper prices fell and, critically, the great ore deposits of Phoenix and the Mother Lode were all but exhausted, the explorations of nearly a decade earlier having been proved correct. The end of the Great War, as it was called, with all its tragedy and social changes, coincided with the end of the ore at Phoenix; it is to the conclusion of that story that we now return.

Closure of the Phoenix and Mother Lode Mines and the Boundary Smelters

Greenwood Smelter Closes…

"Canadian Copper Company [formerly B.C. Copper] General Manager [Hugh] Van Wageenen reports that the Greenwood Smelter was closing because of high operating costs with very low grade of ore from the Mother Lode Mine. Concentrates from Copper Mountain [near Princeton, were not useful because] that ore is not of a self-fluxing nature and hence it is not advisable to build a reverberatory furnace in Greenwood. The Mother Lode mine will be dismantled."

—*The Grand Forks Gazette*, NOVEMBER 22, 1918.

The Last Mine Trains…

"Trains on the Phoenix Branch of the Great Northern Railway will make their last trip today according to an announcement made public by local officials. Such a decision has not come unexpected, however, as officials of the company have gone over the line several times.… It is a direct result of a cave-in at the Granby mines shutting off the Great Northern from ore shipments. The removal of rails from the Great Northern highline at Phoenix has already commenced it is stated."

—*The Grand Forks Gazette,* JUNE 13, 1919.

Proof Against Hoodoo…

"The train crew on the Phoenix 'White Wings' now regards itself as proof against hoodoo. On Friday the 13th instant, they left the terminal at 13 o'clock with 13 cars under train order No. 13 and came through without mishap. Conductor Foote and engineer Tom Needham."

–*The Grand Forks Gazette,* JUNE 27, 1919.

THE MINING AND SMELTING BOOM THAT brought such frenzied activity to Grand Forks, Phoenix and Greenwood in the early 1900s came to a dramatic end soon after the armistice ending the First World War. Copper had been a key strategic material in the all-out war in Europe, but once the fighting stopped, the need for these materials fell dramatically and at the same time the ore reserves were nearly exhausted. In rapid succession the mines closed, at first temporarily and then it became clear that the closures were long-term and probably permanent; with them the smelters closed too. With the ore gone, the mining branch lines soon followed and dismantling was delayed only for formalities and to provide a route to take out the salvage. Almost overnight, it seemed, the mining boom had gone bust.

The B.C. Copper Company's Greenwood smelter closed on November 6, 1918, after several years of intermittent operation due in part to sporadic supplies of ore. The smelter facility was sold for dismantling to Leon Lotzkar who eventually donated the land to the City of Greenwood for a park. Soon only foundations, the tall brick smelter stack and the extensive glistening black slag heaps remained of the once massive smelter.

As ore production at Phoenix came to a halt, the fate of the Grand Forks smelter was sealed. On April 9, 1919, *The Grand Forks Gazette* announced that the Granby Company was abandoning its operations at Phoenix and that dismantling was already underway. Soon after, the crusher at the Granby mine was being scrapped and employees were offered free transportation to move to Anyox, and the paper estimated half of the workforce would accept the offer. Other miners left for work at the new mines at Copper Mountain near Princeton.[1]

In a curious turn of fate, the two major railways

The Granby Smelter in its later years became dwarfed by the growing mountains of granulated smelter slag that rapidly accumulated along the west side of the smelter site. This photo also shows the company offices and residences and other facilities including a tennis court. Such major works required drafting, laboratory, engineering, accounting, purchasing, payroll, maintenance and other administrative and management personnel to oversee the day-to-day functioning of the smelter.
–BOUNDARY MUSEUM.

After nearly 20 years of production, the Granby Smelter at Grand Forks was almost buried in its own slag. The narrow valley in which the smelter was located provided little room for expansion. Had the smelting operations continued for many years, slag would have had to be moved longer distances for disposal. The railways had moved thousands of carloads of Phoenix ore to the smelter and there it had been transformed into bars of copper and mountains of slag. The photo above is from 1914 and the slag heap was over 400 feet (120 m) high. The years of wartime production would add substantially to the piles whose base is formed from the dumped molten slag, like volcanic lava flows, from the earlier years.

–Dave Wilkie Collection, above; Boundary Museum 998.035.008, right.

briefly found themselves in what was almost a race to see who could abandon their Phoenix trackage first, fearing that regulatory bodies might require one line at least to remain in case production resumed. "If we do not get early action, I am afraid the CP will beat us to it," wrote President Ralph Budd of the Great Northern in 1919. However, a cave-in below the Victoria Shaft at Phoenix on April 21, 1919, brought uncertainty to an end when it left both Canadian Pacific and Great Northern tracks hanging in mid-air and cut off the GN from its principal shipping outlet. For GN virtually no traffic remained, reconstruction was unjustified, and in any case the closure of mining and smelting was clearly going to be permanent. Both railways then lost little time in leaving Phoenix behind and, as it turned out, there was no regulatory delay of consequence.[2]

Great Northern services were cut back quickly. On May 12, 1919, the GN hauled its last car of ore from Phoenix to Grand Forks, and to handle the modest extra traffic for a short time the CPR put on an additional ore train. GN passenger services were also reduced to three trains a week, then just three weeks later GN suspended service entirely on its once lucrative branch line, the last passenger train running on June 13. Although the CPR still maintained a daily schedule for a short time, by September service had dropped off to a twice-a-week mixed.[3]

Dismantling of the Great Northern's Phoenix branch and Grand Forks smelter trackage began in the autumn of 1919. A November 7 report noted that two carloads of Doukhobor workers had been brought from Brandon, Manitoba, as labourers for the work. "It apparently being the intention of the railway company to burn their bridges behind them," the *Gazette* noted in a somewhat whimsical turn of phrase, as the dismantling crew started work on the three miles (5 km) of tracks to the Granby Smelter. Although most of the trackage to the smelter was effectively duplicated by the CPR's spur, dismantling got a little ahead of itself when GN workers removed trackage at the smelter yard without obtaining an agreement from Granby, which promptly obtained an injunction to stop the work at the smelter because they had paid for the track and could not effectively remove their machinery without it. Adding to GN's woes, the derrick used to remove the tracks slipped and penetrated the main water supply pipe to the smelter "with a result that water gushed 60 to 70 feet in the air and seriously damaged the roadbed."[4]

In December work was suspended for the winter and

not resumed in earnest until the snows melted from the mountains the next spring. In the interim, the GN crews under John Peterson were employed on bridgework near Republic. In June crews were working again dismantling the Phoenix line and trains were running daily, hauling out rail and other valuable scrap. This work was completed by mid-August 1920. Although in operation for just 15 years, the branch line had carried an impressive and lucrative 7,663,699 tons of copper ore.[5]

Parts of the GN right-of-way were sold to local landowners including some of the Doukhobor communities. In the late 1920s, reports circulated that the GN was considering rebuilding to Phoenix as the outlook for mining improved, but these schemes did not develop. Phoenix remained quiet until the 1950s, when open-pit mining began to recover remaining low-grade ores.[6]

Canadian Pacific retained its branch line to Phoenix for another two years after the GN suspended service, but with no prospects for renewed traffic, dismantling started in late June 1921. Useful equipment, materials and facilities were moved to other locations, including the water tank at Denoro, which was moved to Midway.[7] Equipment from the Emma Mine was salvaged in mid-July and by late August, the Granby Company had moved its crushers and other heavy equipment from the smelter to a point west of the smelter dam so that it would still have rail access once the bridge over the Granby River was dismantled. Late in October the Manitoba Bridge Company began removing the bridge at Smelter Lake and the work was finished on January 26, 1922. Scrap was still being removed the next year. The steel buildings from the Granby Smelter were sold for reassembly at Burke, Idaho.[8]

The once prosperous city of Phoenix quickly faded

A major cave-in under the yards near the Victoria Shaft on April 21, 1919, hastened the end of mining and rail shipments from Phoenix. The three tracks in the foreground in the photo at left belong to the CPR while the two in the background are GN's. The photo is looking northwest. The photo above was taken from the Victoria Shaft ore crusher and shows the No. 2 tunnel ore crusher and ore bins beyond the cave-in. The Granby Hotel is prominent in both views. The collapse cut off the GN from its major shipper. Dwindling ore reserves precluded rebuilding.
–BC Archives, both photos.

The ill-fated Spokane & British Columbia faced not only chronic deficits that doomed the line, but the vagaries of weather that combined with deferred maintenance to cause even more problems. After a derailment in 1915 six miles (10 km) north of Republic, horses substituted for incapacitated Tenwheeler No. 2. By this time the railroad was in deep trouble having lost $1,122,418, but the books had been lost in a 1908 fire. No dividends, apparently, ever were paid. Three engines, 26 freight cars, two passenger cars and two work cars comprised the line's equipment. It was all sold to pay creditors in 1921.[9]

–Boundary Museum, BM987.087.002.1 & 002.2.

away. Some buildings were simply abandoned, but others were dismantled for valuable wood and other materials or were moved to other places. The October 1, 1920, issue of *The Grand Forks Gazette* reported that several hotels had been taken down: the Queens for shipment to Kelowna and the Deane to be relocated to Trail. Other structures were being moved to Grand Forks, and the Granby Hotel had been purchased by a group of Doukhobors, presumably also for salvage. The city hall, the report noted, was the only major building not yet dismantled. On July 15, 1921, a report noted that A. Sercu, sole inhabitant of the former city, had bought the city's remaining chattels.

The end of production at the Mother Lode, the closure of the two railways to Phoenix and the closing of the mines there, along with the shutdown and dismantling of the smelters, had a profound economic impact on the cities of Phoenix, Greenwood and Grand Forks as well as Republic. At Greenwood most of the population quickly drifted away. Grand Forks was better off than the communities that depended almost entirely on mining, because of its agriculture, forestry and role as a regional business centre. The sawmills and fruit packing and processing plants continued to contribute traffic to the railways for many years.

Even before the end of the mining boom the story of the Spokane & British Columbia Railway was coming to an end, and with the closure of the smelters it collapsed into bankruptcy. By early 1918, the newspapers were reporting that it owed depositors $4,570,000 and that it had wrecked the Dominion Permanent Loan Company that had underwritten its construction and operation through its subsidiary, the Grand Forks Townsite Company, which went into liquidation that year and closed on September 27, 1919. The line extended just 36.5 miles (59 km) south of the border and was a long way from reaching Spokane despite claims by W. T. Beck, its president, of having surveys and much of the right-of-way property in hand. The railway was subsequently in the hands of liquidator George T. Clarkson of Toronto. In British Columbia, service continued on the trackage at Grand Forks and on the branch line to Lynch Creek to the north. This line had been built as part of the Kettle River Valley Railway and was subsequently taken over by the CPR as part of its Kettle Valley Railway and operated until 1935.[10]

With little traffic remaining, and the GN still firmly in place to Republic, the S&BC was doomed. *The Grand Forks Gazette* announced on September 26, 1919, that the S&BC was quitting operations the next day, noting that A. B. Buckworth had arrived from Vancouver to go to Republic to close the depot, and that arrangements had been made with the GN to supply motive power for switching at Republic. Not only was traffic all but gone, but the bridges near Curlew were unusable. Abandonment was approved by the Interstate Commerce Commission in April 1921, although it was some while before the tracks were removed. In July the sale of the railway's assets was announced to V. Lamken, railway equipment broker of Seattle: these included about 40 miles (64 km) of steel, the locomotives and some rolling stock.

In just two decades, the boom was over. As Phoenix was being closed down and dismantled, C. M. Campbell, Granby's resident manager, wrote about its impact. From the 16,000,000 tons of Phoenix ores, he reported, "about 350,000,000 pounds of copper and $16,000,000 in gold and silver have been produced, having an aggregate value of $65,000,000."[11] Campbell broke down the figures to produce an interesting profile of the mining and smelting operations. From the $65,000,000, the operating companies earned about $9,000,000 in profits and salvage for property, which cost about $1,000,000, and after an investment of about $4,000,000 for equipment, they were left with a net return after 20 years of operations of $4,000,000.

Campbell's analysis of the expenditures is revealing. The production of the ore, or "ore loaded in the cars," cost $16,000,000 and of this $9,000,000 was paid in wages, with the remainder being consumed in supplies and materials. Over $3,000,000 was spent on explosives from Pacific coast companies; one of the big mines used 80 boxes of dynamite a day during peak

production. Moreover, "fuse and caps from Cornwall, England; mining machinery and machine drills and parts mainly from Quebec and the United States; drilling steel mainly from Sheffield; coal and coke from Fernie; pipe from Quebec and the United States; manganese steel from Hadfield's, Sheffield; rail from Pittsburgh; timber obtained locally; customs duties and freight represent some of the further expenditure of $4,000,000." Railways received $6,000,000 in freight charges for hauling the ore to the smelters. Smelting costs came to about $24,000,000, of which coke and coal were major components. Campbell stated that 2,230,000 tons of coke were consumed. "This means that the entire output of a fair-sized coal mine, with accompanying coke plant, was used to supply the fuel required in smelting Phoenix ores." He estimated the cost of the coke at $10,000,000 and the costs of shipping it to the Boundary District at $6,000,000. Additional costs were in hydroelectric power, and freight to the railways for moving the blister copper to the distant refineries.

The railways—Canadian Pacific, Great Northern and the Kettle Valley Lines—had made the mining and smelting boom possible, and by integrating the industrial and commercial activities of other districts, in particular the Crowsnest, the economic impact of the railways, the mines and the smelters went far beyond the local area. The boom, short though it was, had far-reaching influences. By the time it ended, all but one of the key Canadian Pacific and Great Northern figures had died or retired. James J. Hill and William C. Van Horne died during the First World War, and Thomas Shaughnessey, by then Rt. Hon. Baron Shaughnessey, died at age 79 in December 1923 not long after the last of the Granby Smelter was being scrapped. Only Louis Hill remained active, serving on the GN board until his death in 1948. Jay P. Graves died the same year after an active career based in Spokane that involved land developments, farming and interurban railways. "Granby," wrote that insightful historian of the Inland Empire John Fahey, "towered in Boundary: it employed hundreds in mines and smelter and created jobs for hundreds in the towns, on the railroads, and in the coalfields; it was for a time, the largest copper smelter in the British Empire and second in the world only to the Washoe works at Anaconda, Montana."[12] And that was a fitting epitaph for Granby's presence in the Boundary District.

The Republic Mill, shown at Republic about 1900, was one of the major ore processing facilities in the mining region. Later known as the New Republic Mine, it was the most important mine at Republic with about $1,000,000 produced between 1897 and 1902 before the fortunes of the camp faded. After that production slowed and by 1910 it was averaging from $10,000 to $15,000 a month. Most of its high grade ore went to Tacoma for smelting.
–Dr. Woodrow R. Clevinger Collection, Eastern Washington State Historical Society, L89-183.3.

Above left, Kettle Valley Lines' Assistant Superintendent W. H. Kirkpatrick, who also worked as a conductor, inspects his moribund railway south of Karamin, Washington 1919. Bill Kirkpatrick went on to be a roadmaster with the CPR. The delightful railcar was built by the Cooper brothers at Grand Forks.
–BC Archives, 44190.

Part 3: The Branch Line Years

Great Northern Withdraws Along the Boundary

The Traffic Was Gone...

"There is a good gravel highway from Midway to Bridesville, an important road which almost parallels the branch line as far as Bridesville. The Greyhound Bus line operates a daily passenger and express service between Penticton and Nelson, B.C., passing through Bridesville. A check... of outward traffic from Bridesville shows that from January 1933 to July 1934, there was one carload of rye; two carloads hay; one carload meat. The L.C.L. [less than carload] business consisted of seven shipments of dressed meat and three bundles of wool."

—*Judgement, Application of the Vancouver, Victoria & Eastern Railway and Navigation Company*, Board of Railway Commissioners, Ottawa, December 12, 1934.

"...the abandonment of operation... extending from the international boundary near Midway... to...Bridesville,... is hereby approved."

—S. J. McLean, Assistant Chief Commissioner, Board of Railway Commissioners, December 14, 1934.

In one of the most beautiful portraits of steam railroading in southern British Columbia, Gib Kennedy captured an eastbound *Kettle Valley Express* near Cascade in September 1940 (opposite). Gib Kennedy was a long-time employee of Canadian Pacific and a meticulous researcher and modeller of the Kettle Valley. His articles in *The Model Railroader* magazine advanced the standards of model building during the 1950s and 1960s. His photographs of the Kettle Valley in its last years of steam were some of the finest recorded. This train is powered by P1e 2-8-2 5125 assisted by an N2 3600 for the steep climb ahead. The train includes several coaches, a sleeping car and a café-parlour, which was the best place to enjoy the passing scenery as the train made its leisurely way through the rolling mountains of southern British Columbia.

Great Northern's steam power, like engine 764 above, was wel-maintained even in its last years. Engines had a colourful Rocky herald on their tenders and often had green boilers and red cab roofs. This elderly 4-8-0, is at Kettle Falls between trips to Grand Forks and Nelson on August 14, 1947.

–W. Gibson Kennedy, from a superb print in Lance Camp's collection; and Philip C. Johnson, above.

Operational problems on the Oroville-to-Molson section were a major weakness on the line across southern British Columbia and northern Washington, GN's so-called Third Main Line. The grades and curvature required to traverse the 2,780-foot (nearly 850-m) difference in elevation between the two towns created an ongoing obstacle for crews and continuing excessive costs for management.

As tonnage grew, an operational nightmare evolved. For train crews there were far more dangers running downgrade than climbing up the grade; even with modern airbrakes, the greatest threat was a brake or equipment failure and a runaway train. Westbound, freights easing downgrade were required to stop at Circle and Mount Hull for 10 to 15 minutes to cool wheels overheated by the constant friction of brake shoe pressure. To avoid overheating the driving wheel tires, engineers would set the Johnson bar a couple of notches in reverse position on the quadrant. This setting created compression in the cylinders and it aided in retarding the train and keeping it within safe speed limits. Westbound passenger trains would leave the dining car on the tail of the wye at Midway so that the heavy car would not have to be pulled up Molson Hill on the return trip the next morning. The cooks and waiters stayed with this car overnight and had breakfast ready

A wintertime view from the fireman's side of a Great Northern engine near Myncaster on a cold day in the early 1900s.
–Dave Wilkie Collection.

Firing on Molson Hill...

"I fired a 700 engine on freight out of Oroville to Grand Forks, B.C. Four hundred and twenty-five tons was the tonnage for a 700-class engine, or about 10 or 11 loads we could pull up the hill. If we had good coal we could make it in about two and a half hours, 25 miles [40 km]; if we had poor coal it would take about three or four hours, cleaning fires and blowing up for steam."
– E. T. Stansbury, GN fireman and engineer, 1908–1956.[1]

when the *Spokane Flyer* arrived at 9:38 a.m. the next day, assuming it was on time, to pick up the diner.

Eastbound, freights pulled by the GN's husky and powerful 4-8-0s could handle only 425 tons, and as tonnage grew, extra freights would pull the excess tonnage from Oroville to Molson for later pickup at the summit and movement on by regularly scheduled freight trains. These extras would turn on the wye at Molson and run light or as a "caboose hop" back to Oroville. All of these extra movements and the limited tonnage on the grade contributed to excessive operating costs and argued for the eventual abandonment of Molson Hill.

With the completion of the Oroville–Wenatchee line in 1913, it was only a short time before most eastbound tonnage was taken off the route over Molson Hill and sent south to the main line at Wenatchee. The passenger service, formerly provided by the *Okanogan Owl* westbound, and the *Spokane Flyer* eastbound, was replaced by a mixed train on a triweekly schedule. For some years traffic east of Molson was handled by this second-class train, but by 1931 service between Molson and Oroville ceased and salvage crews took up the rails the following year. Had traffic grown on the route, justifying improvements in the bridges and tracks, it is easy to imagine the Great Northern assigning some of its impressive Mallet locomotives to Molson Hill service, recreating scenes reminiscent of David Moffat's Denver & Salt Lake climbing the Front Range of the Rockies in Colorado. As it turned out, the through route between Vancouver and Spokane never prospered and remained just a group of connected branch lines with little long-distance traffic.

GN officials had discussed abandonment of the Molson route as early as 1917, but at that time any action was judged premature because the farms in the Molson area shipped 610 carloads of grain in 1916–1917 and the town was seen as thriving, at least temporarily. Moreover, GN officials worried that they could not gain permission to break the continuous line provisions of the VV&E charter agreements, and there were complications that would have had to be addressed with the bondholders of the railway. The Great Northern had gone so far as to contact the Kettle Valley's J. J. Warren and made tentative arrangements for the GN to use KVR trackage between Princeton and Midway, if necessary, to maintain the continuous nature of the VV&E.[2] It is interesting to compare the two routes, because it reveals just how different two railways can be between two places. The Canadian Pacific route between Grand Forks and Princeton was 71.6 miles (115.2 km) longer than the Great Northern's trackage and it had substantially longer and more severe grades. Although GN struggled up and down a 2.5 percent grade for the 25 very curving miles (40 km) between Oroville and Molson, the CPR had to deal with 2.2 percent grades on both sides of Penticton, nearly 27 miles (44 km) of 2.2 percent between Penticton and Cedar Creek Summit to the east, and 26 miles (42 km) more 2.2 percent between Princeton and Kirton to the west. Moreover, at Princeton there was 10.7 miles (17.2 km) of such grades, and Eholt was approached from the east by 14 miles (23 km) of 2.3 percent. Between Grand Forks and Midway, the CPR was longer by only 3.7 miles (5.9 km), but the GN line was almost level while the CPR climbed on a steep grade on both sides of the summit at Eholt, making it a much less desirable line from an operational point of view.[3]

In short, Great Northern really did have the edge in distance and operational ease except for the particularly steep climb up to Molson. In any event such considerations became academic because no agreement was put into effect. At a time when the CPR was fully committed to the war effort in Canada and there was a growing crisis in effectively handling burgeoning traffic in the United States, the GN and CPR considered using GN trackage between Grand Forks and Midway, while Kettle Valley tracks would be used between Midway and Princeton, thereby eliminating the costly operations over both Molson Hill and Eholt Summit. Nothing came of the proposals. In 1919 they were discussed again, but the curtailment of operations to Phoenix had both companies wary of any moves that might lock them into maintaining

Conductor Wyatt Eddy spent most of his working life with the Great Northern including many years on the branch lines along the boundary. In traditional conductor's uniform, he took a minute to pose with a passenger in the 1920s photo above. At left, crew members pause beside their locomotive on the GN local between Marcus and Oroville. The scene is at Oroville about 1920 and the 2-6-0, fitted with a big snowplow, is ready for drifts in the hills. The train includes a very elderly combine (coach and baggage car) and a day coach. Engine 480 was originally Spokane Falls & Northern No. 4, built in 1889 by Baldwin (no. 10037). With a tractive effort of just 16,824 pounds, modest passenger trains were its limit. It was retired by Great Northern in 1926 along with three sister engines (SF&N 1–3) and most of the other SF&N engines from the D. C. Corbin years.
–Nancy Denny Collection.

A Great Northern Conductor's Story...

Wyatt Roy Eddy came west from Erie, Pennsylvania, in 1902 to work as a telegrapher on the Great Northern at Bonners Ferry, Idaho, where his salary was $50.00 a month for 12-hour days and no days off, not even Sundays. Soon he became a brakeman and in 1904 was promoted to conductor at Hillyard, on passenger trains running from Spokane to Marcus. In 1909 he was working between Marcus and Republic and in 1911 between Marcus and Nelson. A few years later, in 1916, because the pay was better, he transferred to Whitefish, Montana, as train master, but because the work left little time to be home, he returned to being a conductor on the Marcus–Nelson trains. "I regret your going exceedingly. To me our associations have been very pleasant during the short time I have been in charge of the Kalispell Division," wrote Superintendent H. W. Sheridan in August 1916. "I wish for you the greatest measure of success and happiness in the return to your run on the Marcus Division." In 1918 Wyatt worked on the GN trains to the mining town of Rossland, B.C., and also contracted, but survived, the Spanish flu in the horrible epidemic that killed millions worldwide that year. In the 1920s he was again working between Hillyard and Marcus as a conductor. Retirement from the GN came in 1941 at age 71. When he made his last run on Train No. 701 from Kettle Falls to Spokane, friends turned out at every station along the line to wish the popular conductor well in his retirement.

Next he taught telegraphy in Spokane, where he enjoyed working in the city and "dressing up in a suit and tie." Conductor Eddy died in 1955 after a full career with the Great Northern. His daughter Helen Eddy Charlesworth recalled, when she was 90, that he loved his work, especially being conductor of freights. He didn't like being on the passenger trains as much because there were too many fussy people to deal with, but he always worked hard to keep the passengers happy. He was commended by railway officials on several occasions for his interest in helping passengers with their ticketing problems and in keeping the business on the Great Northern.

–Nancy Charlesworth Denny, from papers and letters in her family collection.

Firing on Molson Hill continued...

"I fired the passenger trains 255 and 256 between Spokane and Oroville. The engines we used were 930, 937, 934, and 925, ... going 25 miles [40 km] an hour with one mail car, baggage car and two day-coaches. With good coal it wasn't bad going, but with poor coal, it was tough going, 25 to 45 minutes late into Molson. From Molson to Midway it was 30 miles [nearly 50 km] downgrade [1.25 %] and the engineer sure would let them roll to make up lost time.

"The speed limit on Molson Hill to Oroville on the passenger was 25 miles [40 km] per hour, on freight it was 10 miles [16 km] an hour. We stopped at Circle City on passenger to cool brakes and wheels. On freight, we stopped at Nine Mile, Circle City and Mount Hull to cool brakes and wheels. If you went any faster you would burn the brake shoes off of the brake beam."

– E. T. Stansbury, GN fireman and engineer, 1908–1956.

services that would no longer pay their way.[4] Moreover, the CPR was hesitant for political reasons to give up a portion of its all-Canadian route across the southern interior for a diversion into Washington, even if it might save substantially in operating costs. Complicating matters, to really make the proposed changes pay, local services to smaller communities in British Columbia would have been curtailed or abandoned. Undoubtedly, too, there were feelings that conditions would improve after the end of the First World War. Companies like the Great Northern had difficulty pulling up trackage just 10 years old without appearing cavalier to local residents and businesses, who often became established on the heels of the railroad, and foolish to residents, stockholders and would-be investors alike. Even though there was little through traffic the communities still used the GN and service was continued through the 1920s. However, with the coming of the Depression it was clear that there was no future for the line. After the First World War, most of the area served by the VV&E, its GN connections in Washington and indeed the CPR in the Boundary and Similkameen districts was struggling as mineral production all but ended in many areas, and logging and sawmilling were then of minor importance.

Permission to abandon had to be obtained from the United States Interstate Commerce Commission and Washington's Public Service Commission. At the same time in Canada, approval had to come from the Board of Railway Commissioners, at least in some circumstances. Dealing with two or more regulatory agencies delayed or complicated the abandonment of Great Northern trackage along the B.C.–Washington boundary but did not prevent it once traffic had declined to trivial levels.

The remaining section between Curlew and Molson, which crossed the border to pass through Midway and Bridesville before recrossing the border again to reach Molson, generated very little traffic except for occasional shipments from the farms and ranches, some express business and mail. On the Canadian side, all freight traffic from the town of Bridesville was being trucked to Rock Creek for transfer to the Kettle Valley Railway. Passenger traffic had all but disappeared. For Bridesville, "Total tonnage inbound and out had dropped from 54,222 tons in 1929 to 3,562 tons five years later. Passenger business had declined from 1,517 tickets in 1929 to a mere 40 tickets in 1933."[5] Late in 1933, permission came for abandonment of the Washington trackage but service was maintained pending the review of the proposals in Canada. On December 13, 1934, *The Penticton Herald* reported the GNR "is still operating and makes a weekly trip from Curlew to Molson. Last Friday there was quite a number of freight cars on. The train usually travels at a rate of 10 mph so as when seen, appears to be moving very carefully and slowly." Approval for abandonment of the railway in Canada between the border near Midway and the crossing back into Washington near Bridesville came on December 14, 1934, and the order authorized the GN to remove the tracks 30 days later. Expensive to maintain and operate because of numerous bridges, including the large curved trestle at Myncaster, and extensive grades and curvature, regular service on the 45.59 miles (73.4 km) of tracks was discontinued on February 27–28, 1935, when the mixed train ran westbound to Bridesville and Molson, and returned the next day. "Sectionmen," noted *The Penticton Herald*, "had been removed to other places." Track removal was delayed until the summer of 1936.[6]

With the abandonment of the trackage between Curlew and Oroville, the railway from Oroville to Princeton and beyond was simply an extension of the GN's branch line northward from Wenatchee through the Okanogan. In any case, this had been the important outlet for nearly all traffic for many years.

To the west, GN service on the railway from the Okanogan to the Similkameen effectively ended at Princeton after 1915 and the opening of the Kettle Valley Railway. The one exception was the unusual service to Tulameen at the southern end of Otter Lake during the winter months. The reason for GN's continuing service to this small community was tied to the prosperous fruit industry of central Washington, connected to the VV&E by the GN branch line constructed north

Abandoned, Oroville to Curlew...

Oroville developed as an agricultural and regional centre just south of the border and retained its train service, but the line east through Molson, Bridesville, Myncaster and Midway to Curlew was too expensive to retain and traffic was all but gone by the late 1920s. The communities suffered a depopulation from a combination of droughts, poor crops, limited opportunities and few resources. The photos show Oroville and its depot; and then a series from August 1934: Bridge 252 at Mile 120.75 looking towards Molson; the water tank at Mount Hull; the sweeping Circle City curve; the station and section house at Myncaster; and at left, the spur to the Chesaw grain elevator and Midway's depot.[7]

–NANCY DENNY COLLECTION, TOP LEFT; OTHERS BY SIMMER, MUSEUM OF SCIENCE & INDUSTRY (SEE ENDNOTE 7).

Ice from Otter Lake...

"At Otter Lake near Tulameen, the GN Railway will cut 3000 cars of ice. This ice cutting plant is one of the largest in western Canada."
–*The Greenwood Ledge*, January 23, 1919.

Officials inspect the ice cutting and loading facilities at Otter Lake on a clear winter day in the 1920s. Cut ice is in the foreground and other blocks are behind the men. Note the ice tongs and picaroons for handling the heavy blocks. Ordinary boxcars were used for the service because during the winter the cold temperatures prevented excessive thawing on the relatively short haul to the ice storage areas. Not all ice went to the fruit industry and some was sold in the communities along the railway. In 1925, 15 carloads were sold to businesses at Princeton for $2.20 a ton.

Facilities included a two-track yard, a loading conveyor and platforms for filling the cars. A bunkhouse, cookhouse and dining room provided accommodations for the 50 to 75 men normally employed, and a barn had room for 16 horses used in the harvest. In 1922 the ice cutters were paid 40 cents an hour with $1.00 a day deducted for board. The next year wages rose to 50 cents, but deductions increased to $1.20.

A single car could carry 30 tons of ice, and as many as 70 cars were hauled in one train although 50 was more typical.

–Okanagan Archive Trust.

from Wenatchee to Oroville. The clear water of Otter Lake, froze deeply, often to two feet (60 cm), provided excellent ice for the hundreds of refrigerator cars that carried Wenatchee apples, Okanogan peaches and other prime fruits and vegetables to markets across the United States. The ice loading plant was completed in November 1915 and cutting was normally done in January and February of each year. Hundreds and sometimes thousands of carloads of ice, representing as much as 50,000 tons of ice, were cut each winter, depending on the weather, demands and quality of the ice. The ice was stored at Oroville, Brewster, Wenatchee, Yakima, Pateros and other major fruit packing, shipping and transit points. In 1922, for example, 60 men worked at Tulameen for about a month to cut the ice. The process was simple and efficient. As much as 50 acres (20 ha) of lake surface was cleared of snow by snowplows and the loading area was cleared of ice. Ideally, the ice was 14–16 inches (35–40 cm) thick, and was scored by an ice plow before men with ice saws cut the thick ice crust into large sections measuring 50 x 100 feet (15 x 30 m). These were floated down to the loading site where they were split into small blocks and pushed onto a conveyor that carried the ice to the cars. A 60-man crew could load 100 cars of ice in a single day. During the ice season one locomotive was assigned to spotting the cars on the Tulameen loading tracks and as many as four others were used to haul the special trains. Because the GN line was generally downgrade to Oroville and other points in the Okanogan Valley, a single locomotive could handle 50 or more cars of ice at a time. Nonetheless, the 1922 pack of 2,300 carloads still required 46 special trains. This was an exceptional ice harvest; previously the highest had been 1,400 cars two years before, and in 1921 only 350 carloads were cut. Until 1924 GN cut the ice itself, but thereafter it was done by its subsidiary the Western

Fruit Express formed in 1923. The winter of 1925 apparently was the last season ice was cut.[8] By that time, mechanical refrigeration plants could produce ice more economically and with more certainty, and the demand for natural ice for the railway disappeared; with it went the only significant traffic for the GN west of Hedley although no immediate abandonment followed.

Floods in 1934 washed out the Similkameen River Bridge near Princeton and it was not rebuilt. The Great Northern already had suspended scheduled service beyond Hedley to Princeton, running there only as needed, although Canadian Pacific continued to operate regular services over the joint trackage as part of the Kettle Valley route. GN moved some fruit shipments as late as 1933 from the Keremeos–Cawston area via Hedley to Princeton.[9] The railway received Interstate Commerce Commission permission to abandon the trackage west of Oroville to the border on October 21, 1935, but permission from the Canadian authorities was not forthcoming. However, in 1937, with little traffic remaining west of Hedley and with approval to abandon the Canadian trackage, the GN removed its 23.5 miles (37.8 km) of line between Hedley and Princeton, and severed forever its connection with the joint trackage west of Princeton.

During this period of low revenues and cost cutting, the notion of sharing trackage between the two railways had surfaced again when Great Northern officials offered to let the CPR use the water-level tracks between Midway and Grand Forks. But the CPR, under Sir Edward Beatty, after some study was not interested, preferring to keep its traffic on the Canadian side of the border. Local traffic complicated the issue at a time when it would have been difficult to abandon any major routes, and as Andrew McCulloch noted in a study he carried out, which is quoted in the adjacent sidebar, the benefits became less than crystal clear. It could be said that political reality trumped engineering efficiency, but that was not new to railways along the Boundary.

The operation of the Coquihalla and Princeton subdivisions created an unusual situation that must have caused GN management considerable annoyance and great expense. The joint trackage agreement retained the GN's rights to extend service through to Hope but at the cost of contributing to the annual maintenance of the Coquihalla trackage. For the CPR, the Great Northern's contribution made the line marginally economic to maintain; the costs of snow removal and of maintaining snow sheds and the seemingly endless trestles and bridges was enormous. Consequently, the CPR was not anxious to lose this de facto subsidy from the Great Northern and was not about to let the GN out of the agreement without a favourable settlement. Finally, over 30 years after the joint trackage agreements between the Great Northern and the Canadian Pacific were made, the arrangements were terminated freeing the GN from its costly obligations. On December 14, 1944, a settlement was concluded between the two railways and their subsidiaries to cancel the original agreement of November 20, 1913, which was to last 999 years, and which had given the VV&E running rights over the Coquihalla. The Great Northern, which technically had taken on all the obligations of the VV&E for 99 years, paid to the CPR $4,455,000 to be rid of the tracks it had used but once. At the same time, GN agreed to pay another $45,000 to release its obligations for the balance of the 999-year term. Concurrently, another agreement provided for the sale of the GN between Princeton and Brookmere to the CPR for $1,500,000. GN's investment in the property was $2,443,000 and it claimed a tax loss of $943,000. In the event this was not possible, the CPR was to pay an additional $400,000. On June 30, 1945, the time limit for the transactions was extended until the first of August the next year.[10] The net cost to the GN was over $2,500,000. For the CPR, another decade and a half of trouble and expense lay ahead before the mountain crossing was abandoned in 1961. The Coquihalla route proved to be a burden for both railways, and they would have been much better off, in the crystal clarity of hindsight, if it had never been built.

Andrew McCulloch Evaluates the GN...

On several occasions CP and GN discussed a consolidation of lines between Grand Forks and Princeton. McCulloch evaluated the idea in 1927. Overall, he saw no clear advantage, as excerpts suggest: "the difference [in distance and operating difficulties] though it may not cut down the number of operating freight divisions should result in a saving for through freight of 30% of the costs as at present. [Better grades would save fuel and] allow heavier loading [which] should bring the total up to 35%.

"After that it is hard to figure any benefit brought in by the change. There can be no amalgamation or combination of the two lines. To take over the GN means abandoning the KVR or adding to the already too great mileage of unproductive railway. If only through business is to be considered the south line is best. As to local business there is not much on either line. [Although] ...on the KVR it does not seem likely that Greenwood or Beaverdell could be abandoned... there is considerable passenger business in and out of Penticton which the KV should not lose..."

–Andrew McCulloch notes and Memorandum, July 26, 1927, courtesy Barrie Sanford.

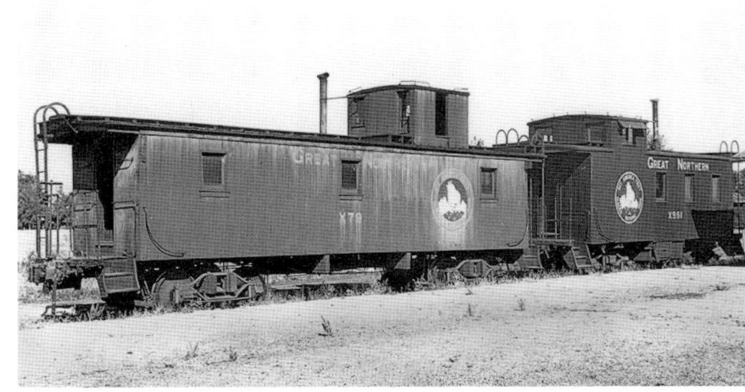

Marcus Was the Division Point for Branch Line Power and Trains to Grand Forks and Points West...

Marcus, overlooking the Columbia River near its confluence with the Kettle, was an important town on the Great Northern in northern Washington. As a division point it saw the comings and goings of trains to Grand Forks and Republic, and to Nelson and Rossland. As mining traffic slowly disappeared, so too did most of the passengers and services declined. Above is F-1 2-8-0 No. 540 with an extra water tank for the run to Republic. At right is the Nelson passenger train, apparently in place of the usual gas car, while above right are two of the cabooses assigned to the freight crews and the rustic old combine, also seen in the train at right, still used on the locals. Roads and automobiles were taking away the life of the trains.
– All photos by Harold Hill, Warren Wing Collection.

Great Northern was an early user of gas- and diesel-powered rail cars or "motor cars." They could pull a coach or several freight cars and were more economical on branch lines than locomotives with conventional trains. GN used them from Spokane to Marcus and from Marcus to Nelson and Republic. "Motor" service started to Republic in 1928, but conventional mixed trains were operated in the 1930s. At left, Train 255, with 1930-built Brill No. 2339, arrives at Marcus from Spokane, while below right, the Nelson gas-electric coasts by the depot. A St. Louis/EMC Model DE-106 gas-electric, 2213 was built in 1925, and made its first trip that December. Below left, GN No. 103, an old freight hog, sits at the engine house. These photos were taken on October 6, 1938.

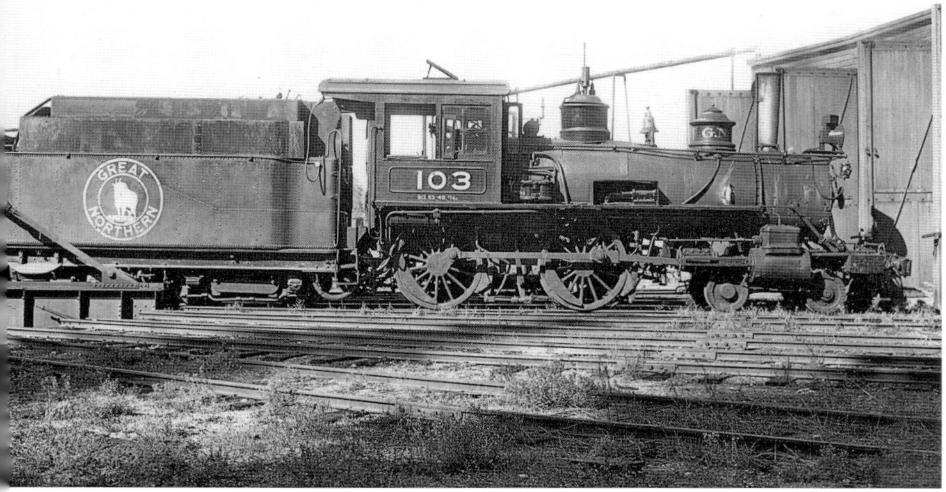

The GNR Moved to Kettle Falls...

When the Grand Coulee Dam was built by the Bureau of Reclamation on the main stream of the Columbia River in the 1930s, it backed up a reservoir of 130 square miles (337 km^2), that was named Franklin D. Roosevelt Lake. It extended almost to British Columbia and flooded the lowlands and wetlands along the river as well as parts of the Great Northern's lines north and east of Colville. A major diversion was built as well as a new crossing of the Columbia. The original site of Marcus was flooded and the division point was moved to Kettle Falls where steam remained in service for a few more years before the arrival of diesels.

–Map by Robert Turner.

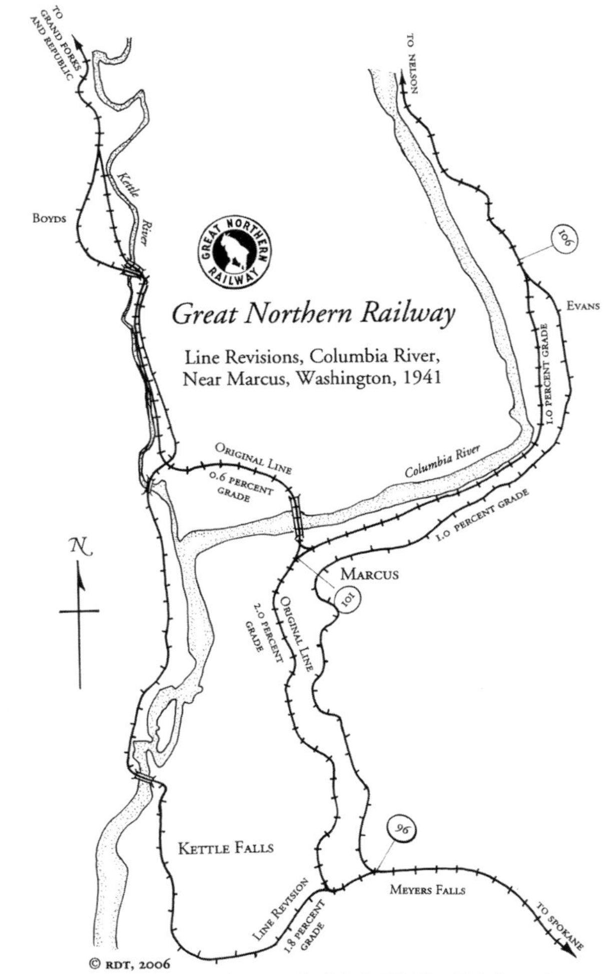

The line revisions were made to accommodate the flooding behind Grand Coulee Dam, on the Columbia River, which created Franklin D. Roosevelt Lake. The dam was built between 1933 and 1942.

SCALE MILES 0 1 2 3 4
KILOMETRES 0 1 2 4 6

The branch lines from Spokane to northern Washington and British Columbia got some of GN's classic steam power in the waning years of steam. GN's No. 722, on the facing page, was a veteran 4-8-0 built by Rogers in 1899, shown at Kettle Falls on August 14, 1947. It worked trains to Nelson and Republic. In just two years it would be sold for scrap. Forty years earlier it was at Phoenix as shown on page 121. Bigger engines, including Mallet 2-6-8-0 No. 1951, ran from Spokane's Hillyard to Kettle Falls. It was photographed, at left, on the same day. Built in 1909 and rebuilt in 1927, it was scrapped in 1950.

–Philip C. Johnson photos.

722
3558

After the Boom Years, Days of CPR Steam

Converting the Steam Locos from Coal to Oil...

"In the year 1949 a large number of locomotives in this territory were converted from coal to oil burning under an agreement with one of the oil companies whereby the price and supply of Bunker oil was guaranteed for a period of three years; also the oil company paid to the Canadian Pacific an amount of $3,500 per locomotive toward the cost of conversion. The guaranteed price of Bunker C oil (not to exceed $1.70 per barrel) assured the railway substantial savings...."

—*Application Study of Diesel-Electric Locomotives for the Kootenay and Kettle Valley Divisions,* Canadian Pacific Railway, 1952.

THE DECADE AFTER THE FIRST WORLD WAR FELT the impact of the closure of the mines and smelters in the Boundary and was followed by another decade whose Great Depression eclipsed previous hard times and gave little respite to the struggling communities on either side of the border. Although by the late 1930s conditions were slowly improving, the sad reality was that it took the outbreak of the Second World War in 1939 to see a substantial revival in the economy. For Canadian Pacific this period was a time when little could be spared for reinvestment in new equipment, especially for secondary services with modest or meagre traffic like those across the southern interior of British Columbia. The beginning of the war virtually precluded new equipment while at the same time imposed increasing demands for service. And as wartime production geared up there was a growing demand for lumber and metals from the resource producing areas of the province. Along the Great Northern's routes to Republic and Hedley, the impact of the war was delayed somewhat because of the later entry of the United States into the conflict, although the economy had increasingly moved onto a wartime footing as the world situation deteriorated.

Unlike the years of the First World War when the Boundary District mines and smelters were contributing critically needed strategic materials to the war effort, during the Second World War the times were quieter in Grand Forks, Greenwood and other towns along the railways. Traffic to and from the Cominco Smelter at Trail became very important during this time. Wartime restrictions on gasoline brought passengers back to the railways and added substantial freight traffic across the system, straining the capacity of aging equipment, still in use because of the depression years, and the railway crews struggled to keep up with demands. Many young railwaymen volunteered, leaving their older friends and very often other family members to work on the railways. At the same time, many women began working for the Canadian Pacific to fill the shortages of personnel.

The war years also focused fear, prejudice and racial tensions. After the Japanese attack on Pearl Harbor and the beginning of the war in the Pacific, Canadians of Japanese heritage were interned and their property seized with little or no compensation. They were forced from their homes on the coast and taken by train to internment camps in southern British Columbia. In the old mining towns of the Boundary and Slocan districts these camps and facilities were hastily developed to house the many people being relocated against their wills. Camps, often with very primitive conditions, or

Grand Forks' Orchards...

"Grand Forks is a town where the irrigated orchards are as well kept as the orange orchards of California. Consequently, and by reason of other favourable conditions, there is a demand in the markets from this place. The river rushes along beside the rails, water seems to be plentiful; but the banks are dry. Logs begin to clog its passage and cover the surface of the water. They are on their way to the mill at Cascade, where the railroad crosses a deep slit in the rock and the water plunges over a high dam built of logs."

–Archie Bell, *Sunset Canada, British Columbia and Beyond,* 1918.

Gib Kennedy captured westbound Train No. 11 drifting down to Cascade behind P1n 5123 and 2-8-0 3659 in 1948. The refrigerator cars and other express business provided important revenue for the passenger services in their later years. Shipping fruit and other perishables as part of the passenger trains moved the high-value products, including apples, peaches and cherries, to markets that were very time and price sensitive. Highways took most of this business in the 1950s.

–Gordon Kennedy Collection.

The railways were a central focus of community life at least through the 1950s when passenger services declined. The assassination on October 29, 1924, of Doukhobor leader Peter V. Verigin, often called "Peter the Lordly," whose funeral train is shown at right, impacted hundreds of people. The train he was on was bombed on Farron Hill and nine people died. The murderers were never found; members of the Sons of Freedom, a radical sect of the normally pacifist Doukhobor community were suspected, but other plots were suggested. The peaceful farming underpinnings of the communes remained, but troubles with authorities continued. In the 1950s, traditionalists did not want their children educated in the British Columbia school programs. Eventually, approximately 200 children were taken from their parents and interned at New Denver between 1953 and 1959 for schooling and allowed only infrequent visits. Lasting bitterness and more protests, demonstrations and violent incidents continued. The CPR cited these events when it curtailed passenger service in the early 1960s. After years of unrest, passions calmed and more flexible education was permitted.
–T. Gushul, Vancouver Public Library, 179.

Members of the Canadian Women's Army Corps relax for a few minutes beside the Midway station, above right, during the Second World War. Thousands of armed forces personnel moved by train during the war.
–Greenwood Museum.

makeshift accommodations in former mining towns, were established at Kaslo, New Denver, Tashme (east of Hope), Rosebery, Slocan City, Lemon Creek, Sandon and Greenwood, and others were operated on the Prairies. Before the war over 22,000 Japanese-Canadians lived in British Columbia, over three quarters of whom were born in the province or had become citizens. Within a few months nearly all, usually moved in passenger trains, were involuntarily relocated to the camps, or semi-abandoned mining towns, in a program that only many years later was recognized by the federal government as a tragedy and great injustice.[1]

By the late 1940s and early 1950s the operational pattern for Canadian Pacific in the Boundary area was characterized by three types of services: passenger, through freight, and way freight or local trains. Passenger services were generally limited to the daily trains, the westbound *Kootenay Express* (Train No. 11) and its eastbound counterpart the *Kettle Valley Express* (Train No. 12), which ran between Vancouver and Medicine Hat. These trains each averaged about 10 cars daily although there was variation in requirements. Helpers were usually needed between Labrathe and Farron on trains of eight or more cars, and between Grand Forks and Eholt on trains with six or more cars. In the other direction, nine-car trains needed helpers from Midway to Eholt and trains with six or more cars needed assistance between Cascade and Farron.

Through freight trains averaged about one and a half a day on the Boundary Subdivision. Typical freight trains over the Boundary Subdivision handled 22 cars east and 23 cars west between Nelson and Grand Forks and 25 cars east and 29 cars west over the railway between Grand Forks and Midway. Tonnages were usually 1,300 eastbound and 1,375 westbound between Midway and Grand Forks, while between Grand Forks and Nelson tonnage was a little under 1,100 tons each way. Helpers were needed for the freights over Farron and Eholt summits. Crews were based at Farron and assisted either eastbound or westbound trains as needed, and other crews were stationed at Grand Forks for trains to Midway.

Over Eholt Summit, eastbound trains over 725 tons ran with helpers all the way from Midway to Grand Forks. Westbound trains with less than 550 tons could manage with a single 2-8-0. On heavier trains, one or two helpers were used. Sometimes, trains were doubled between Grand Forks and Eholt with either one or two locomotives, meaning that the engines would take as much tonnage as they could to Eholt and return light to Grand Forks to pick up the remaining cars and take them to Eholt on a second trip. From there, the cars were reassembled into a longer train and taken downgrade to

Double-headed steam power was the norm on passenger and freight trains on the Boundary Subdivision as these Farron Hill photos show.
–W. G. Kennedy, Gordon Kennedy Collection, at left; above, Philip R. Hastings, California State Railroad Museum, 202.

Midway where they were turned over to Kettle Valley Division crews for the journey west.

With few local industries aside from fruit packing and sawmilling at Grand Forks, there was little local traffic. A main line way freight operated between Grand Forks and Nelson once a week in each direction, picking up and distributing freight cars as needed to local shippers and businesses. If traffic warranted, more trains ran as needed. Fresh fruit was often handled on the passenger trains to improve speed of delivery and reduce spoilage.

The picture that emerges of the Boundary Subdivision between Nelson and Midway is one of a rugged railway, difficult and costly to operate, that sets the stage for its dieselization, and despite optimistic traffic projections in the immediate post–Second World War years, its eventual abandonment some decades later. The mountain grades took their toll and, although not as infamous as the Coquihalla's, were even more demanding of the railway's resources. On an annual basis, more double heading of locomotives on the daily passenger

Ruling Grades, CPR's Boundary Subdivision

Total Miles, Average Ascending & Maximum Ascending Grade[2]

Eastward

Midway to Eholt: 17.9 miles (28.9 km), 1.25%, 2.3%
Cascade to Farron: 23.8 miles (38.3 km), 1.91%, 2.4%

Westward

Labrathe to Farron: 25.6 miles (41.2 km), 1.90%, 2.4%
Cascade to Billings: 1.0 miles (1.6 km), 1.23%, 2.1%
Grand Forks to Eholt: 13.9 miles (22.4 km), 1.83%, 2.4%

Kootenay Division Steam Locomotives...

In 1952 the CPR assigned the following classes and numbers of locomotives to the Kootenay Division:[3]

Passenger Engines		
G3f-g	4-6-2	3
P1n	2-8-2	3
Freight & Helper Engines		
D10g	4-6-0	2
M4g	2-8-0	1
N2a	2-8-0	20
N2b	2-8-0	9
N2c	2-8-0	3
P1n	2-8-2	7
R3b-c	2-10-0	2
V4a	0-8-0	1
Yard Engines		
M4a-g	2-8-0	5
N2a	2-8-0	1
R3c	2-10-0	1
V4a	0-8-0	2
V5a	0-8-0	2

Philip R. Hastings, one of the most celebrated photographers of North American steam, visited Nelson and the Boundary District in 1951, not long before the division was dieselized. He captured this steamy view of Nelson with a lineup of power including P1n 2-8-2 5261 and 2-10-0 5758 as well as other engines being serviced or awaiting the next assignment. Engine 5261 was rebuilt from CPR 2-8-0 No. 3667 in 1949 and was retired in September 1959.

–California State Railroad Museum, 213.

trains was required on the Boundary Subdivision than on the Kettle Valley Division's Coquihalla, Merritt and Princeton subdivisions collectively. In fact, the Boundary Subdivision's total for the use of double headers was exceeded only when the Carmi Subdivision's totals are included with the other three subdivisions just mentioned. Costly and unproductive in revenue terms, light engine movements were the highest of any division in the entire southern route across British Columbia and Alberta. In efficiency terms, freight operations along the Boundary route were challenging because of the many mountain summits to be crossed. On the Boundary Subdivision in 1950 locomotives operated nearly 74,000 miles (119 000 km) as helpers and double headers on through freight trains, and over 35,000 miles (56 000 km) were added as light engine movements. Canadian Pacific concluded that tonnage over Farron Hill was "the controlling factor in its main line operation between Nelson and Ruby Creek."

In a postwar climate of cost cutting, coupled with expanding needs for new equipment, these conclusions led inevitably to dieselization and a cloud over the future of its lightly used southern British Columbia lines.

Canadian Pacific Steam Along the Boundary...

CPR steam locomotives had a family appearance that was often characterized by Elesco feedwater heaters across the smokeboxes of larger engines. At left, 5101, a P1d 2-8-2 on Train No. 11, on May 25, 1942, at Cascade, and right, the 2534, a G2s 4-6-2 of 1908 vintage at Castlegar, gets attention from its engineer. Engine 5157, in side view, was another well travelled P1 2-8-2 with characteristic CPR cab and tender, photographed at Castlegar, while veteran 3677, an N2 2-8-0, heads the *Kootenay Express* west out of Robson West in 1938, and 3676 pulls into Castlegar, lower left and 3626 steams quietly at Grand Forks. The 3600 N2 class 2-8-0s were common on Canadian Pacific's Boundary Subdivision and just before dieselization there were 32 of these tough old engines at work. Most Kootenay locomotives were coal fired and the 3600s lacked stokers, so firemen had their work cut out for them, but in the last years of steam, oil burners were installed on many engines. L. S. Morrison, a young operator with the CPR, loved to photograph the steam power at Castlegar and Coykendahl and captured 5157, 3677 and 3676, while Gib Kennedy, also a CPR operator, recorded 5101 and 2534.

–The photo of 3626 at Grand Forks and these prints are from Lance Camp's collection, with thanks to Mo Morrison and Gordon Kennedy.

1804
1804

Diesel Years and Decline

Diesels Brought Efficiencies and Widespread Change...

"The elimination of helper operations, the reduction in elapsed time—diesel vs. steam—and the fewer number of trains that will be operated make impracticable the present practice of splitting the Boundary Subdivision into two train runs, with crews from Nelson operating as far as Grand Forks and another set of crews stationed at Grand Forks for the run between Grand Forks and Midway. With diesels, crews will be stationed at Nelson to operate through to Midway thus doing away with Grand Forks as a crew terminal with a considerable saving in expense to the Company."

—*Application Study of Diesel-Electric Locomotives for the Kootenay and Kettle Valley Divisions,* Canadian Pacific Railway, 1952.

Conventional Passenger Trains Were Dying...

"Heavy losses in operating conventional passenger trains on the Kettle Valley line, may force their replacement with popular 'streamliner' type self-propelled rail diesel cars.... The new type trains, which can be operated as single cars or in tandem, already are proving highly popular in other parts of Canada."

—E. C. MacPherson, Canadian Pacific Railway, *Press Release*, 1957.

A new face came to the CPR with the arrival of streamlined diesels in the early 1950s. These are Canadian Locomotive Company demonstrators at Penticton in September 1951.
–Harry Davis photo, Okanagan Archive Trust.

By the late 1940s it was clear that the days of steam on the railways of North America were drawing to a close. Postponed by the Second World War and at the same time hastened by technological developments during those years, dieselization was inevitable, based on economic savings that could see a railway recoup its investment in diesels in just a few years, and at the same time improve speed and customer service. Less romantic, but modern-looking and colourful, diesels could outperform even the newest steam locomotives, in nearly any application. It was no wonder that the 30- and 40-year-old machines, long overdue for replacement but used on most branch lines, were prime targets for retirement in the face of new diesels. Maintenance was dramatically reduced, frequent water stops and other steam servicing facilities were not needed, fewer locomotives were required and often crews could be eliminated.

The Great Northern introduced diesels to its main line, branch line and yard operations in substantial numbers during the early 1940s, but the Second World War intervened and further dieselization was restricted. When limits on locomotive purchases were removed, GN, it seemed, could not buy diesels fast enough. The railway even eliminated its electrification through the Cascade Mountains by substituting diesels in the mid-1950s. GN was also a leader in the postwar years in introducing new diesel-powered streamlined trains to its transcontinental *Empire Builder,* the *International*s that ran between Seattle and Vancouver, and other services including the *Red River* that ran to Winnipeg. With far less traffic and a limited number of trains on what remained of the old VV&E and related lines in Washington, GN used mostly roadswitchers, including Alco RS2s and RS3s and General Motors GP7s and GP9s, on the old mining branch lines. Later these were replaced when newer types of power were acquired for main line services, and, in turn, more up-to-date engines were available for the branch lines.

Mining traffic, meanwhile, had dwindled and all but disappeared with forest products taking over as the main traffic on the railway from Colville to Republic. Rail service fluctuated to meet the business offered, but

Photo, Facing Page...
Crossing the Columbia River was a major obstacle for the Great Northern when it built west from Marcus to the Boundary and Republic in the early 1900s. However, flooding after the construction of Grand Coulee Dam between 1933 and 1942 required a major line revision, shown on the map on page 178. A new bridge over the river was opened in 1941 and remains in use. At left, a Burlington Northern freight with three Geeps, as the General Motors roadswitchers are called, accelerates as it nears the eastern end of the span and heads towards Kettle Falls from Grand Forks and Republic in summer 1978. From Kettle Falls another train will take these cars and others south to Spokane.
–Robert D. Turner.

Brand-new Great Northern diesels, 214 and 213, painted in the GN's beautiful orange and green Empire Builder colour scheme, arrived at Wenatchee's Appleyard from Alco for use on wayfreights and other services, including the branch lines to British Columbia. It is September 25, 1949, and the RS2 diesels arrived just the day before. General Motors' FT road diesels, the type of locomotive that helped persuade GN to scrap its steam fleet, are in the background.

At right, GN 204, another Alco RS2, but built in 1947 and fitted with a steam boiler for passenger service, was still working branch line freights on Washington branch lines in 1961. In 1964 it was traded in to General Electric and its two younger sisters followed the next year.

–Dave Wilkie, above; Robert D. Turner at right.

remained adequate for the continuation of the operation of the branch line. On May 2, 1970, James J. Hill's Great Northern Railway became part of a new railroad, the Burlington Northern, when it was merged with the Northern Pacific and the Burlington (Chicago, Burlington & Quincy), along with the Spokane, Portland & Seattle (which was jointly owned by the GN and the NP) and other subsidiary railroads to create one of the largest rail companies in the United States. Such a system had been the dream of Hill in the early 1900s but he had been unsuccessful in the regulatory environment of his time to achieve the integration of the three major properties in which he had interests. Although the merger brought an end to the Great Northern as an independent property, it certainly achieved Hill's goals and was part of a series of major mergers that reshaped the face of American railroading in the late 1900s.

The use of second-generation diesels on the Republic line became more frequent in the post-merger era. Trains operated from Colville to Republic and return. At Colville larger trains were assembled combining the branch line traffic with shipments to and from Colville and Kettle Falls and the former GN branch lines to Nelson before being moved south to Spokane.[1]

For the CPR, overall dieselization in the postwar years was a little slower than it had been for the Great Northern, but the mountainous divisions of British Columbia were a priority. After successfully replacing all steam on the CPR's Esquimalt & Nanaimo Railway on Vancouver Island in 1949, attention shifted to the east and the dieselization of much of the traffic on the Wells River line that provided an important link between Montreal and Boston. Next came the very important Schreiber Division on the main line between Cartier (just west of Sudbury) and Fort William along the north shore of Lake Superior. The railway then turned its resources to the main line between Revelstoke and Calgary and then the Kettle Valley and Kootenay divisions across southern British Columbia. With their high operating costs and frequent use of helper locomotives, these divisions were a natural target for the economies of diesel operation.

When dieselization of the Kettle Valley and Kootenay divisions was being considered by the CPR in the early 1950s, 92 steam locomotives were in service on the two divisions, and all but a handful predated the First World War in their construction, although they had been modernized. Further efficiencies came after nearly

Canadian Pacific and Great Northern used different locomotives on their Boundary lines. CP purchased Canadian Locomotive Company's C-Liners for the Kettle Valley and Kootenay divisions. Above, a C-Liner is assisting an RDC at Penticton in 1960. Above left, GN used General Motors SD9s to Republic as shown near Curlew on March 29, 1967. At left, C-Liners accelerate east of Greenwood and through all but abandoned Eholt on August 10, 1967, then are crossing the Kettle near Cascade a week later, accompanied by roadswitchers.
–Dave Wilkie.

CONDENSED SCHEDULE

READ DOWN ♥45 Dayliner Daily	MEDICINE HAT - NELSON - TRAIL - PENTICTON - VANCOUVER			READ UP ♥46 Dayliner Daily
A.M.		Mountain Time		P.M.
10.15	Lv	‖ MEDICINE HAT ‖	Ar	8.30
11.07		Bow Island		7.33
11.45		Taber		6.56
12.10		Coaldale		6.32
12.25	Ar	‖ LETHBRIDGE ‖	Lv	6.20
12.50	Lv	‖ LETHBRIDGE ‖	Ar	5.40
1.24	Lv	FORT MACLEOD	Lv	5.05
2.03		Pincher		4.28
2.37		Hillcrest		3.55
2.45		Blairmore		3.49
2.50		Coleman		3.44
3.10	Ar	Crowsnest, B.C., MT	Lv	3.30
2.25	Lv	Crowsnest, B.C., PT	Ar	2.10
2.57		Michel		1.40
3.01		Natal		1.36
3.33		Fernie		1.06
5.10	Ar	CRANBROOK+	Lv	11.35
5.25	Lv	CRANBROOK	Ar	11.20
6.31	Lv	Yahk	Lv	10.16
7.18		Erickson		9.32
7.26		Creston		9.24
9.50	Ar	‖ NELSON ‖ PT	Lv	7.00
10.00	BUS Lv	NELSON PT	Ar BUS	7.00
11.10	BUS	Castlegar	BUS	5.40
11.50	BUS Ar	Trail	Lv BUS	5.00
P.M.				A.M.
♥45 Dayliner Mon. & Thur.				♥46 Dayliner Tues. & Fri.
A.M.				P.M.
7.00	Lv	‖ NELSON ‖ PT	Ar	4.15
7.50	Ar	Castlegar+	Lv	3.27
c 8.10	BUS Lv	Castlegar+	Ar BUS	e 3.25
c 8.50	BUS Ar	Trail	Lv BUS	e 2.45
7.50	Lv	Castlegar+	Ar	3.27
10.09		Grand Forks		1.06
10.55		GREENWOOD		12.20
11.15	Ar	Midway	Lv	12.00
11.20	Lv	Midway	Ar	11.55
11.55		Westbridge		11.21
12.27		Beaverdell		10.49
c 3.20	Ar	PENTICTON+	Lv	e 8.00
P.M.		(CONTINUED ON NEXT PAGE)		A.M.

–Robert Turner Collection.

all passenger engines and all but two freight engines had been converted to cleaner and more efficient oil fuel from coal. "Over half the steam locomotives on the Kootenay and Kettle Valley territory are of the N2 Class, all of which are over 37 years of age. They are generally not desirable for service on other than mountain territory and, in view of this and their age, it is considered they would largely serve as yard and spare power and any surplus... would be scrapped," noted the CPR's study of dieselization for the district.[2] The undeniable conclusion for Canadian Pacific was that steam was obsolete, however romantic it might be, and the economics for the railway were absolutely clear.

The savings from dieselization were many-faceted. For the Crowsnest to Ruby Creek operations, savings in fuel costs were projected at $1,395,000 a year or a decrease of 56 percent. Savings from wages for enginemen were likely to be over $415,000, reflecting a decrease of 33 percent, and from trainmen's wages additional savings of $286,000 were forecast, for an overall reduction of 47 percent. Diesels did not require firemen to operate, and although it took some time for new agreements with the railway unions to be negotiated and for these savings to take effect, they were certainly significant.

Although the investments in diesels were substantial, studies leading to the change estimated that in overall terms Canadian Pacific could reduce its costs by 34 percent. Operating expenses were estimated to fall by $2,852,000. The overall return on investment was at least 15 percent. There were other savings as well that were not calculated specifically. These came from substantial improvements in the speed of trains over the divisions (in part derived from shorter terminal times), the increased utilization of freight cars because less time was taken in transit, and the diesels' dynamic brakes which saved brake shoe and wheel wear in the mountainous territories as well as improving train safety and efficiency.[3]

Of course the impact on employees was substantial. Helper terminals disappeared from Grand Forks and overall fewer crews were required to move the tonnage. Maintenance staff also declined as facilities were removed and centralized, and some employees and their families had to relocate. The traditional job of fireman faded away as new work rules took effect. Men also needed training for the new equipment, but the new diesels were certainly cleaner and easier to work. Fortunately, the need for fewer crews was offset somewhat by increasing traffic in the prosperous post–Second World War years.

At that time there were three main builders of diesel-electric locomotives in Canada producing locos to essentially the same designs as their American parent or affiliated companies: Montreal Locomotive Works (MLW), building American Locomotive Works locomotives; the Canadian Locomotive Company (CLC), making Fairbanks-Morse designs; and General Motors Diesel (GMD), building locomotives developed by its United States counterpart, the Electro Motive Division of General Motors. Initially, Canadian Pacific bought locomotives from all three builders for use on the Kootenay and Kettle Valley divisions. These included 10 GMD and 2 MLW cab or A-units, as well as 2 MLW B-units, all fitted with steam generators for passenger service; 6 A-units from CLC and 10 from MLW and 14 B-units (some from each builder) for freight service; 14 GMD roadswitchers; and 10 switchers for yard duties from MLW and a further 6 from GMD. Most of these locomotives were delivered in 1953.

However, within a few years of entering service the CLC units from other divisions were concentrated on the Kootenay and Kettle Valley divisions, and most diesels of other types being used in southern British Columbia were transferred elsewhere. This facilitated multiple-unit operations and the concentration of the CLC-Fairbanks-Morse locomotives on these two divisions simplified maintenance, crewing and operations. At the same time, even though it was never part of management decisions, it gave the motive power on the CPR's southern B.C. lines a distinctive appearance. Maintenance was based at Nelson where a large new diesel servicing facility was built. The FMs became inextricably linked with the

The CPR's RDCs, or *Dayliners*, were modern, comfortable and quiet. Air conditioning was a decided plus in the hot southern interior summers. But for all their modernity, they could not retain enough passengers to keep the service on the Kootenay and Kettle Valley divisions from being terminated. Mail contracts had been taken over by trucks and highways offered more convenience and independence for travellers. These quiet scenes, with RDC 9107 on Train No. 45 on August 3, 1961, capture the charm and the dilemma of the service: there were simply too few passengers to keep the *Dayliners* running. In the last years, service was down to two trains a week each way between Nelson and Penticton, although the service was daily from Penticton to Vancouver and from Medicine Hat to Nelson. At left the train is at Grand Forks and above at Midway. Midway's C&W station survived the abandonment of the railway to become the home for the Kettle River Museum.
–Philip C. Johnson.

railway through southern B.C., and most were based at Nelson until 1975 when they were retired.[4]

When Canadian Pacific began its dieselization of the Kootenay and Kettle Valley divisions, locomotive-hauled conventional passenger trains were still in service and diesels, equipped with steam heating equipment, took over from steam locomotives on the *Kootenay Express* and *Kettle Valley Express* as well as other passenger assignments. However, this was a brief interlude of transition. Improvements to roads and the development of air services were eroding passenger bookings on the railway and Canadian Pacific looked for a more cost-effective alternative to its existing trains. Before the completion of the Hope–Princeton Highway in 1949, road travel across the southern interior was difficult at best and time-consuming, but the improvements that followed in the next few years, including extensive paving, made travel by automobile much more feasible and pleasant. Moreover, less-than-carload freight, express and mail could be moved faster and more cheaply by road, particularly over short to medium distances, than by train, and in 1957 trucks took over these services.[5]

The interim cost-effective solution Canadian Pacific found was the stainless steel Rail Diesel Car, or RDC,

A First Generation of Diesels...

Diesels on the Great Northern and Canadian Pacific brought colour if not as much drama to the railways. Burlington Northern absorbed the GN, including its lines to northern Washington, on March 1, 1970, but it took some time for the merger to remove the Great Northern's Rocky, the mountain goat, emblem from its equipment. The engine and caboose of a GN wayfreight at Oroville, on a very cold but crystal clear January day in 1971, was a case in point. On the facing page, GN GP9 roadswitchers in orange and green and the newer blue and white colours switch a few cars at Republic in July 1969, in pre-merger days. Their contemporaries on Canadian Pacific in southern British Columbia were Canadian Locomotive Company (Fairbanks-Morse) H16-44 roadswitchers and C-Liners that were serviced at the Nelson diesel shops, shown in September 1974, the year before they were retired. –Robert D. Turner, five photos.

684

4104
Canadian Pacific
CP Rail

4105
CANADIAN PACIFIC

8714
Canadian Pacific

Canadian Pacific's CLC, Fairbanks-Morse diesels may not have been very beautiful, but they were robust and functional with a certain brick-like charm, and gave two decades of hard service in the mountains. These photos show two H16-44 roadswitchers and a C-Liner, hard at work on a freight from Midway, above left, through Grand Forks to Farron Hill in 1974. The C-Liners could be used only as leading or trailing units with the roadswitchers because they lacked multiple-unit connections at the front.

–Robert D. Turner, four photos.

developed by the Budd Company of Philadelphia. These were self-propelled, air-conditioned cars that came in several configurations, and Canadian Pacific bought them in substantial numbers in the mid- to late 1950s for its branch line passenger services. However, a shortage of the new RDCs forced Canadian Pacific to operate short conventional trains consisting of an engine, one baggage car and a coach, between October 1957 and February 1958 after the discontinuance of the *Kettle Valley Express* and *Kootenay Express*. Unfortunately, not even the new RDCs, or *Dayliners* as they were called by Canadian Pacific, could reverse the decline in passenger traffic and revenues. Service along the Boundary was reduced to just twice weekly, and by 1962 Canadian Pacific revenues were only one seventh its expenses for the passenger trains. The company was given permission to abandon services in January 1964. During this period, protest bombings attributed to the Sons of Freedom Doukhobor sect had disrupted rail service, and this situation, the CPR argued, threatened the safety of passengers and contributed to its case for discontinuance.[6]

The 1960s brought another dramatic change to the once important Coast-to-Kootenay railway route: the closure of the Coquihalla Pass route between Brookmere and Hope. The advantages of its direct route through the mountains were largely negated by its high maintenance and unreliability in winter and made the financial arguments undeniable, especially when the longer but much easier route via Merritt and Spences Bridge was available. Winter storms hit the Coquihalla Pass in November 1959 resulting in several washouts, closing the railway for the winter. But a review of the situation and the possible traffic over the route led Canadian Pacific to apply to abandon the fabled railway through the mountains. In July 1961 approval came for dismantling the railway and the work of removing the tracks followed. The Coast-to-Kootenay railway was not severed, but at best it was certainly severely bent by the detour via Spences Bridge. Unfortunately, the closing of the Coquihalla line was but the first domino to fall over the next years.

Although Canadian Pacific had invested considerable money in modernizing its Kootenay and Kettle Valley divisions in the 1950s and early 1960s, it became increasingly clear that the southern divisions were not going to be an effective alternative or bypass to the main line and the heavy traffic in coal, grain and other bulk commodities that was developing. The main line was systematically improved, and because of their mountainous nature, with many pusher grades, and the lower standards of construction that restricted the size and weight of engines and equipment, the two southern divisions effectively became regional branch lines, surviving only on local traffic that was fading away. A major impact of this decision came in 1962 when no

Similkameen Ending...

Keremeos, upper left in September 1967, was cut off from the GN when high water washed out the bridge over the Similkameen, six miles (nearly 10 km) north of the border, in 1972, shown at top. By 1970 the reduction works at Hedley were a ruin soon to be demolished. Trackless roadbed and sections of bridges survived for years including parts of Bridge 330 near Nighthawk, B.C., in July 1994. The enclosed bridge near Keremeos, above, used for a road and shown in the late 1960s, was rebuilt in 2004.

–Bridge 330 and washed-out bridge, Dave Wilkie; others Robert D. Turner.

through westbound freight was sent over Farron Hill from Castlegar and other points to the east. Instead, all of this traffic was routed to Cranbrook and then sent north over the Windermere Subdivision, which had been upgraded, to Golden and then westward over the main line. This effectively removed all but local traffic from the Boundary Subdivision and also reduced traffic over the Kettle Valley and in particular over the mountainous Carmi Subdivision.

The expensive-to-maintain mountain crossings were a severe liability as the years went by, and the next one to fall to the inexorable economics facing the railways was the CPR's Carmi Subdivision between Midway and Penticton. This was part of the original Kettle Valley, and service was discontinued in 1973 except for a few trains running west to Beaverdell from Midway. Abandonment followed in 1978 and the tracks were removed over the next months. With this closure of the railway through the spectacular Myra Canyon, the Coast-to-Kootenay railway was only a memory. The surviving trackage between Penticton and Spences Bridge in the west, and from Midway east to Castlegar and from there south to Trail, or further east to Nelson, were just long branch lines, often with low traffic.

A Second Generation of Diesels...

In a June 1978 scene at Nelson, above, modern General Motors SD40-2 locomotives have replaced the old CLCs and are outside the diesel shops, and the yard is full of 40-foot and 50-foot boxcars. Nearly all of this activity disappeared after the diesel shops closed in 1988, followed by the loss of the Boundary traffic, and as other shipments dwindled. In 1988 the division headquarters were closed and shifted to Revelstoke. Reductions in traffic through Nelson destined for Trail came in December 2001 with the closure of the Sullivan Mine at Kimberley. The railway was fading away. Along the Boundary Subdivision operations continued over the mountains to Grand Forks and Midway until late 1990. On the facing page, lower left photo, on December 6, 1982, Extra 4218 West was at Mile 46.5 on its way to Midway, high above Lower Arrow Lake. The Montreal Locomotive Works C-424 leading was an unusual visitor to the Kootenays. Switching continued at Grand Forks, as shown on October 26, 1990, after the rails over Farron were removed. –Robert D. Turner photos, except Extra 4218 by John Leeming, at right.

The Columbia River bridge, shown when new on page 52, is remarkably unchanged after over 100 years of service. It still has its original masonry piers, and the swing span built to permit sternwheeler navigation, although no longer used, remains in place. This train is handling shipments for Trail and the Celgar pulp mill in February 1997. The detail above shows the rollers, now dry of grease, for the swing span at the top of the circular pier. Many decades have elapsed since the last sternwheeler passed through.
–Robert D. Turner.

Great Northern and its descendants on GN's Seventh Subdivision to Grand Forks and Republic. Above left, crossing Curlew Lake on August 9, 1967; at left, Burlington Northern power at Kettle Falls with combined trains from Spokane in 1982; and above at Curlew as a BNSF crew switches lumber cars in August 2004. By the mid-1980s, service to Republic had ended and the lumber transfer at Curlew was the normal end of track until abandonment south of Danville in 2006. –DAVE WILKIE ABOVE LEFT; OTHERS, ROBERT D. TURNER.

Canadian Pacific freights still roll along the Kootenay River and cross on the original bridge over the Columbia, but all traffic is headed either for Castlegar and its pulp mill, or south to Trail and its smelter, the last in the entire region. This scene shows a westbound CPR freight passing the Brilliant hydroelectric dam on the Kootenay River in February 1997. With its falls and rapids, the Kootenay was unnavigable and this led to the construction of the railway to bypass the river in the 1890s, before it was developed for electric power generation. –Robert D. Turner.

The Celgar pulp mill, now owned by Zellstoff Celgar, has been an important source of traffic at Castlegar since its opening in 1960 by the Celanese Corporation of America. Between 1989 and 1993 the mill was substantially rebuilt. The mill is located just upstream of Robson West from where sternwheelers used to ply the Arrow Lakes before dams were built on the Columbia River. The mill and a CPR switcher are shown above.

At right, a Kettle Falls International train arrives at Grand Forks in deepening late afternoon shadows on October 11, 2006, a few months after the railway ceased operating to San Poil.

–Robert D. Turner, above; Bob Wetham, two photos at right.

BNSF transferred its Wenatchee-to-Oroville trackage to RailAmerica Incorporated for operation under the very appropriately named subsidiary, Cascade & Columbia River Railroad Company.[10]

To the east there were changes too. On October 29, 1997, BNSF sold the tracks between Salmo and Troup (tracks that had once connected through to Nelson) to Canadian Rail Track Materials, Inc., which later removed the rails and other materials. Then in 1998, B.C.-based shortline operator International Rail Road Systems (owned by International Reload Systems) acquired from BNSF its remaining trackage north of Columbia Gardens, just above the B.C.–Washington border. BNSF retained about four miles (over 6 km) of trackage south of Columbia Gardens on the Canadian side of the border. The new company provides terminal and switching services for lumber shippers and for the concentrates moving to the Trail smelter. Later that year the trackage north of Parks (siding) to Salmo was abandoned. A yard and truck interchange terminal was maintained at Parks for shipments to and from the Trail smelter.[11] Although BNSF sold the tracks, it retained the right-of-way between Columbia Gardens and Salmo.

To the west, traffic levels on the BN to and from Grand Forks, and lumber shipments along the line to the south in Ferry County, remained moderate although service beyond Curlew was limited, and the trackage beyond San Poil to Republic was removed in 1983. Late in 2004, OmniTRAX, a successful shortline operator, took over the BNSF trackage north of Kettle Falls, Washington, including the line to Columbia Gardens, via a lease, and the trackage to Grand Forks and San Poil, through a purchase agreement.[12] This new railway was named the Kettle Falls International Railway, LLC. OmniTRAX, an affiliate of the Broe Companies, Inc., is based in Denver, Colorado, where the companies have their headquarters, and is the largest privately owned rail service company in North America. It also operates the former CPR trackage between Sicamous, on the CPR main line through British Columbia, and Vernon in the Okanagan Valley under the name Okanagan Valley Railway, as well as many other railroads in North America through lease, operating or purchase arrangements. However, in October 2005, Pope & Talbot, the major lumber producer in the area, announced the closure of their mill at Midway, which shipped lumber through the reload near Curlew, and the consolidation of its operations at Grand Forks. Subsequently, in July 2006, Kettle Falls International filed to abandon 28.5 miles (45.5 km) of track from just south of the border near Danville to near milepost 77 at San Poil.[13] One more piece of the railway network to the mines and smelters, developed by James J. Hill, came to an end, but to have survived over 100 years was no small accomplishment.

Postscript: The Boundary Railways in Retrospect

After 100 years, all that remained of the once extensive Canadian Pacific and Great Northern trackage in the Boundary region was just a few short crossings of the border operated by successors to the Great Northern that served resource-related industries exporting lumber and wood products, as well as residual mining materials and, to the east, processed metals from the Trail smelter. James J. Hill might well have had some satisfaction in seeing that some of the GN's branch lines outlived the CPR's extensions along the boundary.

In the end, the Great Northern expansion into southern British Columbia and northern Washington was not a profitable venture overall, but parts of it were, and it is as a network of branch lines that it is best envisioned. Rival Canadian Pacific trackage did not always fare much better. The Coquihalla—the direct route to the coast for both railways—was never a financial success in an overall perspective, and some other CPR branch lines faded quickly. However, it is too easy to credit the decisions leading to construction simply to the rivalry between the two great railway companies and to say that the expansion into these areas was a mistake.

If in retrospect James J. Hill with his Great Northern and William C. Van Horne and Thomas Shaughnessy with their Canadian Pacific were too quick to build into remote, lightly populated territories, they were in their own time certainly criticized for being too slow to expand. If they overbuilt trackage on the speculative development of mines and smelters, or on the assumption that immigration would continue to new heights, so too did the backers of the Pacific Extension of the Milwaukee Road (completed in 1909 and later electrified in a burst of further capital investment), the Pacific Great Eastern, the Canadian Northern, the Grand Trunk Pacific and other railways in pre–First World War western North America. The last years of the 19th century and the decade and a half before the beginning of the First World War were seductive times. Immigration to the Northwest states and British Columbia from Europe and from the eastern seaboard of the United States, mining development at centres such as Rossland, Phoenix and Republic, and the settlement of the vast agricultural lands of the Northwest states and the Canadian Prairies all seemed destined for endless expansion. And where one boom town faltered and faded, another seemed to rise to take its place. Railways were the key to the success or failure of nearly all of these ventures.

James J. Hill's personality, his immense energy and drive, coupled with his push for reciprocal free trade between Canada and the United States, were also keys in his approach to branch line construction. His frequent speeches and support for free trade during the early 1900s coincided with Prime Minister Wilfrid Laurier's push for reciprocity. The defeat of Laurier's Liberal government in 1911 must have been a blow to Hill because it spelled the end of his hopes for free trade. Had reciprocity–free trade been enacted by the two countries, Hill's Great Northern would have been in an enviable position to capitalize on the changes in markets and traffic patterns that likely would have followed.

Railway construction to Phoenix, Grand Forks and Republic was clearly justified. The mines were highly productive and showed promise of extensive reserves of ore, and the nearby smelters were an early and profitable traffic bonanza. Politically, the CPR probably could not have avoided construction and the Great Northern would have been foolish to ignore the traffic potential. Within a few years, Phoenix was producing the third highest car-loading statistics in British Columbia. When the mines were worked out, the rail lines were withdrawn

The British Columbia Provincial Museum's Museum Train visited communities along the Boundary Subdivision in 1977, powered by locomotive 3716. It was the last time a steam locomotive would ever climb over Farron Hill and reach Grand Forks, Greenwood or Midway. The next year BC Rail's Royal Hudson 2860 made a tour to promote the bicentennial of the explorations of Captain James Cook in 1978 and ran as far west as Castlegar with its exhibit train.

–Robert D. Turner

A century after the Great Northern and Canadian Pacific battled for the traffic along the international boundary, only remnants of their once extensive operations remained. At left former CPR engine 6703, attractively painted and lettered Grand Forks Railroad Co., provides switching services to several industries at Grand Forks. Engineer Shelley Dahl and brakeman Dennis Ghon are continuing a long tradition in August 2004. In the scenes above, also from August 2004, a Burlington Northern Santa Fe freight blasts into Laurier from Kettle Falls, while above, later the same day, the train returns from Curlew. –Robert D. Turner.

quickly but their short duration does not mean they were either unnecessary or unsuccessful. Such rapid abandonments reflect a pragmatic view of railway construction that suggests the tracks should not survive after the major sources of traffic have disappeared.

West of Midway, the GN did not rush construction; indeed the pace was if anything very leisurely, working just fast enough to demonstrate some progress and commitment, yet minimizing costs. Because it had major projects underway elsewhere, it built at a slow pace that required over five years to reach Princeton and another five to see the route to the west completed. This was clearly not a frantic pace pushed by unthinking competition. Certainly competition and leverage with the CPR had its role in the GN's construction, but Hill was taking his time and letting things develop. Perhaps it is best to view these expansions into the Okanogan, Similkameen and other districts along the international border as speculative investments in a frontier area. Similarly, the CPR paused at Midway, in no rush to build westward and connect the Columbia & Western with its main line far to the west across the mountains.

The beginning of the war in Europe in 1914 and Canada's mobilization changed everything in the west. Immigration to Canada from Europe ceased almost entirely, many men volunteered and left for the Western Front, farms were left and often abandoned, and mine production, after an initial lull, suddenly became strategically important. In the Northwest states, the settlement boom continued for another year or two but soon the war had its inevitable impact as the United States was cut off from Europe and in 1917 was drawn into the widening war.

The war disrupted immigration and settlement schemes and brought a financial crisis to several major railway systems. After the First World War, automobiles, depleted ore deposits and changed perspectives on the future all had an impact on the viability of the prewar railway schemes. By the early 1920s, bankruptcy overtook the Milwaukee, the Pacific Great Eastern, the Canadian Northern and the Grand Trunk Pacific. However, both Great Northern and Canadian Pacific remained solvent through economic recessions and depressions and the difficult wartime years, despite the sometimes less than successful financial performance of some of the Boundary District branch lines. Prudent management certainly had its influence on their decisions over expansion and abandonments.

A railroad builder who anticipated a boom and built a profitable line to foster and encourage the boom is seen as a clairvoyant, a business genius. But one who built a railroad in anticipation of a boom that faded too quickly and lost money is too easily labelled a person of lesser skill and ability. More often than not, the difference was one of forces beyond the abilities of anyone to foresee. Some might even call it luck.

It is also too easy, with the passing of nearly a century, to compress the series of events and decisions that led to the boom years of railway construction along the Washington and British Columbia boundary. What appear to be closely connected events were not always as much a part of a long-term plan as a series of incremental decisions in a shifting economic and political climate. It is perhaps best to consider the Great Northern's operations along the boundary as a series of individual, but connected, enterprises that need to be evaluated separately as well as in the context of overall developments. Individually some branch lines were successful or at least had a sound business basis for construction. Others were tied into charters that required construction of unprofitable sections so that work could proceed on sections that could make money. Still others were speculative ventures that may have been ill-considered, or simply were built at the wrong time. And of course there was nationalism. The Canadian Pacific had been built as a Canadian transcontinental railway at great expense avoiding cost saving detours through the United States. GN was unfettered by these concerns, but the CPR was not, and it certainly influenced later decisions about line sharing or amalgamation along the international boundary.

Hill's Canadian branch lines were not a single entity. Some sections—Curlew to Oroville, the Kootenay

Heritage Memories...

The railway story has been well preserved in many places on both sides of the border. Stations, ranging from the large wooden building at Nelson from 1902 that served as headquarters for the Kootenay Division, shown above, to more modest small-town stations or depots survive. Castlegar, Grand Forks, Midway and Oroville all have preserved stations. Princeton's also survives although much modified. Many miles of right-of-way have been saved and developed as trails and other areas accessible less formally. The large bridges near Cascade have been rebuilt for trail use, and the Castlegar bridge is still used by the CPR. Many heritage buildings are valued in communities all along the now long-gone railways. Reminders of the smelters and mines are less pleasing perhaps, but nonetheless fascinating, although the sites are potentially hazardous. The smelter stack at Greenwood is a very tangible reminder of the once busy industrial complex.

–Robert D. Turner.

In years gone by when the steam railways crisscrossed the Boundary District, hundreds of people made their careers keeping the trains runnings reliably and safely. There were small stations all along the routes, often in very isolated locations, where an operator, often alone, relayed train orders giving instructions to the train crews. Section crews patrolled the lines and carried out repairs or removed boulders, logs or debris from the tracks. L. S. Morrison, "Mo" to his many friends, spent his working life with the CPR including assignments at Castlegar and Coykendahl before moving to the Okanagan. For a young man it was a lonely life in the mountains in winter at remote stations like Coykendahl or Farron. But positions were usually determined by seniority and with time opportunities came for a change of location and promotion.
–L. S. Morrison.

Railway & Navigation lines north of Bonners Ferry, the VV&E Fraser Valley lines and the trackage west of Princeton—were products of excess expansion and corporate positioning in the event of free trade, and never justified their costs with revenues, although some were intertwined with a larger picture that dictated, or at least was perceived to be linked to, charters and corporate guarantees. The little used and very expensive extension to Hope was justified only in terms of providing the VV&E with a continuous route across southern British Columbia. For the CPR, construction through the Coquihalla was certainly in this category. Other trackage, however, is not so easily dismissed. One of the two routes to Phoenix was probably unnecessary, but these branches were not major construction projects and they did produce considerable revenue. The railways may not have liked duplicate trackage, but shippers often appreciated the competition and lower freight rates. Other branch lines proved productive and long-lived. The trackage from Marcus to Grand Forks and Republic approaches a century of operation; the core of the GN's Spokane Falls & Northern/Nelson & Fort Sheppard trackage is still operational. The former GN coastal line between Seattle and Vancouver, whose history deserves a detailed examination on its own, is a major route connecting the important urban centres and seaports in the region. The purchase or construction of these routes was clearly not in error. Any railroad that lasts for 90 or 100 years was certainly not a mistake in the longer term.

We may never know how seriously Hill considered the connected branch lines as a through route. If he thought of them that way, it was probably an error, and a very expensive one at that, but it is more likely that beyond the bluster and hyperbole of speech making he was more pragmatic. Privately, he may well have considered the enterprise more a collection of related industrial branch lines, stitched together for expediency and because of the availability of the VV&E charter in an era when free trade was a serious possibility and apparently limitless expansion seemed a reality. The connections that were uneconomic were the price of the branch lines that held more promise and often proved financially viable. Indeed, as David C. Jones points out in a thoughtful consideration of GN's dilemma over the Canadian branch lines, the slow pace of abandonment of these unproductive lines was due in large measure to the GN's reluctance to apply for abandonment of sections of this "continuous" route and risk violating the VV&E charter. GN officials realized that there could be difficulties gaining permission to abandon these sections, knowing the unprofitable lines would be a serious financial burden. Moreover, because the lines ran across the international boundary, as Jones further points out, sometimes there was uncertainty over who really had authority to authorize abandonment.[1]

Hill's trackage to the Crowsnest mines, which like the coast line deserves a treatment of its own, provided the GN with a source of high-quality coal in the west and served the railway for two decades before the railway began to convert most of its Lines West locomotives to oil burners. At the same time, the excellent grades on the GN's Crow's Nest Southern gave it an efficiency unattainable by the CPR. The coal also supplied the smelters and other industries along the Great Northern encouraging the development of other sources of traffic. Unlike the CPR's Crowsnest Pass line, which was a multi-purpose route, this was essentially a resource railway, and once it had served its purpose, it was removed.

The most lasting legacy of these railways may have little to do with the mines and the smelters that drew them into the southern interior of British Columbia and the mining districts of northern Washington. It will reside in the scenic and historic qualities of the abandoned grades across the region, and their associated heritage sites. These sites are now in many areas parts of the Trans Canada Trail and other recreational trails that have become major revenue-generating attractions and sources of enjoyment for residents and visitors to the region.

The saga of these railways illustrates the way things that seem so permanent, so indestructible and so tangible, can be so transient. The boom years in the Boundary District were short, less than 25 years in fact, and far less than a working lifetime. Many miles of the railways, and the mines and smelters they served, flourished only between the late 1890s and the late teens of the next century. By the 1920s, the huge smelters were demolished, towns were abandoned, workers were gone, railways were dismantled. Only ruins and overgrown trails remained. We expect some things to be ephemeral and short-lived, but railways and smelters seem to be assets that are permanent and indestructible. But they are not. Few of our major engineering works survive long beyond their immediate usefulness, and it is always surprising how quickly industrial processes, modes of transport or other focal points of industry and commerce are made obsolete or, for resource industries, are depleted. With them go the capitalists, investors, workers, managers, merchants and everyone else who for a short time drew a living from these enterprises. Usually little is left, certainly not a legacy of wealthy, or even sustaining, communities. Sometimes the most visible legacy is just abandoned foundations and slag dumps. Fortunately, there are exceptions along both sides of the border including Grand Forks and the smaller communities of Midway, Greenwood, Republic and Curlew that struggled on long after the boom years. It is to the credit of less flamboyant enterprise, often people working on their fruit ranches, on farms or in the forests and mills, and more recently in other enterprises including tourism, that a newer economy has emerged and sustained the towns and villages.

The industrial past of the Boundary District had many qualities, and is at best a difficult era to understand. It was certainly dynamic and dramatic, peopled by larger than life figures struggling for wealth and power. Enormous challenges were faced and overcome, and all with an excitement and drama hard to imagine as much as a century later. At times it was carefree and an era of little regulation or restraint, at times appealing at least in retrospect, like long lost youth. When better to be a young engineer going out to challenge the mountains, build a railway and find fame and fortune? But lest this romantic image obscure the other side of the story, it is fundamental and essential to recall that it was also highly dangerous for the railway workers, miners and smelter workers at a time when these workers and their families were often little more than a commodity in the production process and received little compensation for their hard labour. Racial and cultural prejudice was endemic and widespread. There was little opportunity, for example, for a young Asian immigrant or a young woman to seek a new or different life or to use a talent to the fullest. Safety standards were minimal and equipment was often crude and unreliable. The industries were enormously polluting and destructive of water, forests, land and air, leaving a legacy that survived long after the wealth generated locally had long gone, and the profits of investors and speculators had been dispersed or invested elsewhere. There was no thought, for example, of restoring landscapes or cleaning toxins from waste dumps or streams. It was a frontier economy, often dramatic, romantic and exciting, driven by a quest for wealth and for some, a secure future. On balance, the picture is not always an appealing one, yet it has a fascination and a complexity that draws one back in imagination and through images long past to understand and to contemplate, and perhaps to consider some lessons for the future. It has taken generations for directions, attitudes and prejudices to change, and even now, I wonder, how much different would things be if a new Phoenix were found. How much better at managing our future would we be than were those industrialists, political leaders and workers of the not so distant past when immediacy and necessity presided over legacy?

On a positive note the establishment of trails over the once busy railway grades is a wonderful use for these scenic and historic linkages through the Boundary. I like to think the engineers and labourers who worked so hard to build these railways would be pleased to see thousands of people still travelling along their trails to the copper camps of the Boundary.

Crossing the international boundary at Laurier, a BNSF train accelerates into British Columbia on its way to Grand Forks and Curlew in August 2004, a little over 100 years after the railway was opened. It was no coincidence that this border crossing of the Great Northern was named for Liberal Prime Minister Sir Wilfrid Laurier who supported free trade, or reciprocity, between Canada and the United States when James J. Hill sought traffic for the Great Northern in the mining districts of British Columbia.

Below, a Kettle Falls International train switches at Grand Forks in October 2006; a last link with the Great Northern.

–Robert D. Turner, above; Bob Whetham, below.

Notes

Additional references are included in the sidebar citations or are indicated in the text.

Introduction, pages 14-19.

[1] The name Crowsnest was written Crow's Nest (and sometimes Crows Nest) in many early schedules and descriptions. It officially became "Crowsnest" in 1902 but the simplified form did not get wide use for a long time. Businesses, such as the Crow's Nest Pass Coal Company, often retained earlier spellings.

[2] The biographical notes and references to Van Horne, Shaughnessy, and James J. and Louis W. Hill were based primarily on Lamb, 1975, *History of the Canadian Pacific Railway*; Cruise and Griffin, 1988, *Lords of the Line, The Men Who Built the CPR*; Knowles, 2004, *From Telegrapher to Titan*; Martin, 1977, *James J. Hill and the Opening of the Northwest*; Hidy et al., 1988, *The Great Northern Railway, A History*; and Malone, 1996, *James J. Hill, Empire Builder of the Northwest.* See also the biographies of Van Horne (d. 1915) and Shaughnessy (d. 1923) both by Theodore D. Regehr, and Van Horne (d. 1916) by David G. Burley in the *Dictionary of Canadian Biography.*

Railway Names, Abbreviations and Charters, pages 20-23.

[1] For details of the Canadian charters see Robert Dorman, 1938, *A Statutory History of Steam and Electric Railways in Canada.* We have also referred to most provincial and federal legislation in *Statutes of British Columbia,* or *Statutes of Canada,* and to Cail, 1974, *Land, Man, and the Law,* Hidy et al., 1988, *The Great Northern Railway, A History,* and Middleton and Keyes, 1980, "Corporate List, The Great Northern Railway Company," *History,* No. 143, pp. 8–19. Websites include: www.railamerica.com and www.omnitrax.com.

Canadian Pacific and Great Northern Build to the Boundary, pages 24-29.

[1] Schwantes, 1993, *Railroad Signatures Across the Pacific Northwest,* p. 17.

[2] Ibid., particularly Chapter 2, pp. 57–80.

[3] See Harris, 1997, *The Resettlement of British Columbia.*

[4] Hidy et al., 1988, *The Great Northern Railway, A History.*

[5] Turner and Wilkie, 1994, *The Skyline Limited, The Kaslo and Slocan Railway,* p. 22, and Sanford, 2004, *Royal Metal,* p. 16. The New Westminster Southern Railway charter was authorized in Chapter 27, SBC, 1883, which lapsed, then was reinstituted by Chapter 36, SBC, 1887, with subsequent amendments.

[6] Veazey, Phyllis, 1983, "John Hendry and the Vancouver, Westminster and Yukon Railway: 'It Would Put Us on Easy Street.'" *BC Studies,* No. 59, p. 53, notes that in Sept. 1907 when Hendry ended his GN association, he received for his personal services, $125,000 for the VW&Y, $47,000 for the VV&E and $13,500 for the Victoria Terminal Railway & Ferry Company, another GN subsidiary. My thanks to Dr. Frank Leonard for letting me read a draft of a paper about John Hendry and his relationship with the Great Northern.

[7] Hidy et al., 1988, p. 77, Veazey, ibid., pp. 44–63, and Sanford, 2004, *Royal Metal.*

[8] The provincial legislation was *An Act to Incorporate the Vancouver, Victoria and Eastern Railway and Navigation Company,* Chapter 75, SBC, 1897, and the federal legislation, *An Act Respecting the Vancouver, Victoria and Eastern Railway and Navigation Company,* 61 Victoria, Chapter 89, assented to June 13, 1898.

[9] Turner and Wilkie, 1994, *The Skyline Limited, The Kaslo and Slocan Railway,* p. 22; Sanford, 2004, *Royal Metal,* p. 16. The New Westminster Southern Railway charter was authorized in Chapter 27, SBC, 1883, which lapsed, then was reinstituted by Chapter 36, SBC, 1887, with subsequent amendments. *The Railway and Shipping World,* June 1898, p. 100.

[10] Ibid., August 1898, pp. 151–152.

[11] Regehr, 1976, *The Canadian Northern Railway,* pp. 62–63; Eagle, 1989, *The Canadian Pacific Railway and the Development of Western Canada,* in particular for the details of the political background to the VV&E charter and its evolution, pp. 41–42, 109–112; and Fleming, 1991, *The Railway King of Canada,* pp. 79–80.

[12] The Soo Line's full name was the Minneapolis St. Paul & Sault Ste. Marie Railway.

[13] Hidy et al., 1988, *The Great Northern Railway, A History,* pp. 97–98.

[14] See Lamb, 1975, *History of the Canadian Pacific Railway,* pp. 202–203 and 225–227. Also Cruise and Griffiths, 1988, *Lords of the Line.* I saw "Von Horne" Ave [sic] in Cuba in 1998.

[15] The Shaughnessy and Van Horne correspondence quoted here and elsewhere is from their papers in CPR Archives, Montreal.

Phoenix and the Mountain of Copper, pages 30-53.

[1] Different measures sometimes were used in reporting. Tons used in some of the federal Department of Mines reports equalled 2,240 lbs., very close to one metric tonne, while other "tons" equalled 2,000 lbs. (a short ton), which is equivalent to 0.9072 tonnes. Unfortunately in many cases the actual measure of a ton is not reported. To avoid compounding errors and confusion, I have left the tonnages as originally reported, without metric conversion.

[2] See John Fahey's excellent biography, 1994, *Shaping Spokane, Jay P. Graves and His Times.*

[3] Ibid., p. 14.

[4] Fahey, 1965, *Inland Empire, D.C. Corbin and Spokane.*

[5] Ibid.

[6] Kirk and Alexander, 1990, *Exploring Washington's Past,* pp. 30–32. See also the official website of the Colville Confederated Tribes, www.colvilletribes.com. Today, the Colville Reservation includes 1.4 million acres (565 000 ha) mostly in Okanogan and Ferry counties, about half of its original size.

[7] Quoted by Fahey in *Inland Empire,* p. 188, citing *The Spokane-Review,* April 21, 1898.

[8] Turner, 1984, *Sternwheelers and Steam Tugs,* and Turner and Wilkie, 1994, *The Skyline Limited, The Kaslo and Slocan Railway.*

[9] The discussion about Robson and Castlegar is based on CPR timetables and CPR *Boundary Sub, Condensed Profile,* revised to 1915. Mileages also varied. In 1899, for example, the distance between Castlegar (before being called Castlegar Jct.) and West Robson was shown as 1.6 miles. Mileages were taken from West Robson until the Columbia River bridge was completed, and then Castlegar Jct. became the starting point for the trackage to Trail and Rossland, as shown in the 1902 *Time Table No. 1* for the Pacific Division. Mileages west to Midway were determined from Nelson because the entire route was part of the Boundary Section. A spur track ran from near the east end of the Castlegar bridge down to the Columbia.

[10] *The Boundary Creek Times,* January 29, 1898.

[11] *The Nelson Miner,* March 5, 1898, reported that Heinze retained his other holdings in B.C. that included a sawmill at Trail, logging camps along the Columbia, some mining properties, townsites and one half of the land grant received to that point. See also McNelis, 1968, *Copper Kings at War;* Lamb, 1977, *History of the Canadian Pacific;* and Eagle, 1989, *The Canadian Pacific Railway and the Development of Western Canada.* An agreement dated 27 May 1897 was signed between Heinze, Herbert S. Holt and Charles S. Hyman, and Van Horne and Shaughnessy, requiring Heinze and associates to secure a federal charter for the C&W and to build and lease the line to the CPR. Shaughnessy papers, CPR Archives.

[12] *Statutes of Canada,* 1898, C. 61 and Cail, 1974, *Land, Man, and the Law.* See particularly pp. 159–168, and SBC Victoria, 1896, Chapter 54.

[13] *An Act Respecting the Repurchase by the Crown of Certain Railway Subsidy Lands,* SBC George, 1912, Chapter 37.

[14] For a biography of Tye see American Society of Civil Engineer, *Memoir 208,* "William Francis Tye, Died January 8, 1932," by John G. Sullivan and F. P. Gutelius. Copy courtesy Barrie Sanford.

[15] One board foot is equivalent to a piece of lumber 1x12x12 inches (2.54x30.5x30.5 cm).

[16] McRae Pass is located between Castlegar and Christina Lake to the north of Highway 3 over Blueberry Summit. To CPR crews, the railway route was known as Farron Hill.

[17] *The Nelson Miner*, September 2 and 13, 1898.

[18] See Turner, *West of the Great Divide*, 1986, chapter 2, and *The SS Moyie, Memories of the Oldest Sternwheeler*, 1991, pp. 14–15. Details of the tunnel equipment are from *The Railway and Shipping World*, December 1898, p. 267.

[19] Turner and Wilkie, 1994, *The Skyline Limited.*

[20] Fahey, 1965, *Inland Empire, D. C. Corbin and Spokane.* Later, when the Great Northern extended its trackage to Grand Forks, Phoenix and Republic, and the operations were consolidated, the trackage was operated along with the lines to Nelson and Rossland as the Spokane Falls & Northern Division of the Great Northern and former SF&N locomotives, and others from the GN, were used throughout the division.

[21] The finished tunnel appears to have been about 2,991 feet (911.7 m) in length, the difference probably being made by excavations at the portals. The elevation of the summit of the switchbacks was 3,648 feet (1112 m) and the tunnel 3,214 feet (979.6 m), a difference of 434 feet (132 m).

[22] *The Cascade Record*, August 12 and 19, 1899. Mackie's subcontract was for the bridge spans.

[23] Ibid.

[24] Richard Armstrong to T. P. Coffee, August 22, 1899.

[25] *The Railway and Shipping World*, August 1900, p. 225, and also see note 25.

[26] Smith's report was published in *Sessional Papers*, 1900, "Columbia & Western Railway Report, On the Works of Construction on the West Robson–Grand Forks Division as Performed up to the 21st. Day of September, 1899," pp. 394–402. For Shaughnessy's comments, see *The Railway and Shipping World*, October 1899, p. 292.

[27] Armstrong to Coffee, August 23, 1899.

[28] J. G. Sullivan to Andrew McCulloch, March 2, 1937.

[29] The report also was published in *The Railway and Shipping World*, August 1900, pp. 225–229.

[30] *The Boundary Creek Times*, September 16, 1899. Columbia & Western Ry. Co., *Time Bill No.1 and Special Rules,* September 18, 1899.

[31] *The Boundary Creek Times*, October 21, 1899.

[32] Ibid., October 28, 1899.

[33] Ibid.

[34] Report of Henry B. Smith in *Sessional Papers,* 1900, "Columbia & Western Railway, Report on the Works of Construction on the Grand Forks-Greenwood Division, as Performed up to 22nd. Day of November, 1899," pp. 403–410. See also *The Railway and Shipping World*, September 1900, pp. 222–224.

[35] Kennedy, 1986, *Canadian Pacific in Southern British Columbia, The Boundary Subdivision*, p. 3.

[36] *The Kootenaian* (Kaslo), November 30, 1899, and December 14, 1899. *The Cascade Record*, December 9, 1899, reported that 170 guests were accommodated.

[37] *The Railway and Shipping World*, June 1900, p. 173.

[38] *The Grand Forks Daily Gazette*, March 1, 1900. The name of this newspaper changed a number of times in the early 1900s, but *The Grand Forks Gazette* was the most commonly used.

[39] Ibid., January 6 and May 19, 1900.

[40] *The Grand Forks Daily Gazette*, May 23, June 2 and July 14, 1900.

[41] Ibid., August 25 and September 1, 1900. For a detailed description of the smelter see C. A. Bramble, 1901, "The Boundary Creek District. The Town of Grand Forks and the Granby Smelter," *The British Columbia Mining Record*, Vol. 7(6), pp. 207–222.

[42] Mines Branch, 1907–1908, *Report on the Mining and Metallurgical Industries of Canada.* Department of Mines, Ottawa, Ontario, p. 86. (The British Columbia sections of this extensive study apparently were written by R. R. Hedley.)

[43] Based on site visits and reports in *The Railway and Shipping World*, November 1900, p. 330, and December 1901, p. 362.

More Railways for Grand Forks and Republic, pages 54-67.

[1] The seven trustees were Clarence McCuaig of Montreal, Warren Miller of New York, John C. Ralston of Spokane, Albert Azro Ayer of Montreal and R. G. Edwards-Leckie, D. F. Hallahan and A. F. Burleigh all of Republic. Ralston was a civil engineer and in charge of the surveys for the line.

[2] Based on a full account in *The Daily Colonist* (Victoria), September 1, 1900.

[3] Sanford, 2004, *Royal Metal, The People, Times and Trains of New Westminster Bridge*, provides a detailed account of the construction and impact of this structure and its railway connections.

[4] Some reports refer to Porter Bros. & McArthur as the contractor for bridges.

[5] *The Phoenix Pioneer*, August 31, 1901.

[6] The following account is based on a lengthy report in *The Grand Forks Miner-Gazette*, January 11, 1902, and other reports.

[7] Ibid.

[8] *The Grand Forks Miner-Gazette*, April 19, 1902.

[9] Ibid., May 3, 1902.

[10] Ibid., May 10, 1902.

[11] *The Victoria Times*, August 4, 1902, p. 2.

[12] *The Midway Dispatch*, November 17, 1902.

[13] Howland Bancroft, 1914, *The Ore Deposits of Northeastern Washington, including, the Republic Mining District* by Waldemar Lindgren and Howland Bancroft, p. 175. This paper also includes details of the Belcher Mining District in general.

Phoenix, the Mines and the Smelters, pages 68-79.

[1] *Phoenix Pioneer and Boundary Mining Journal*, clipping "The City of Phoenix, Premier Mining Camp of the Dominion of Canada," 1906, pp. 27–28c.

[2] Campbell, "The Passing of a Canadian City," *The Canadian Mining Journal*, April 29, 1921, p. 339.

[3] William Yolen Williams was replaced as superintendent at Granby by A. B. W. Hodges, the smelter manager, and was appointed by William Nichols. Williams became a consultant. He was given a lavish farewell dinner in Phoenix. See Fahey, 1994, *Shaping Spokane, Jay P. Graves and His Times*, p. 77.

[4] Campbell, 1921, "The Passing of a Canadian City," p. 335.

[5] Fahey, 1994, *Shaping Spokane, Jay P. Graves and His Times*, pp. 82–83. For a detailed discussion of the union movement in southern B.C., but particularly Rossland, see Jeremy Mouat, 1995, *Roaring Days, Rossland's Mines and the History of British Columbia.*

The Phoenix and Motherlode Mines, pages 80-91.

[1] LeRoy, 1912, *The Geology and Ore Deposits of Phoenix, Boundary District, British Columbia*, pp. 55 and 73.

[2] Mines Branch, 1907–1908, *Report on the Mining and Metallurgical Industries of Canada.* Department of Mines, Ottawa, Ontario, pp. 128–131. Our descriptions are adapted from the detailed discussions in this extensive paper, and from information from Peter Corley-Smith, Research Associate, RBCM, who as a young man worked with similar mining techniques in African copper and gold mines.

[3] Joe Smuin, personal communication, 2005. For steam shovels, see George E. Cole, "Steam Shovels in Mines," *The Mining Record*, Vol. 10, No. 12, December 1903, pp. 886–889.

[4] *The Grand Forks Daily Gazette*, November 14, 1903.

[5] Mines Branch, 1907–1908, *Report on the Mining and Metallurgical Industries of Canada*, pp. 147 and 156.

[6] LeRoy, 1912, *The Geology and Ore Deposits of Phoenix, Boundary District, British Columbia*, pp. 86–94.

[7] *The Grand Forks Gazette*, March 3, 1917.

The Boundary District Smelters, pages 92-111.

[1] Peters, 1895, *Modern Copper Smelting.*

[2] LeRoy, 1912, *The Geology and Ore Deposits of Phoenix, Boundary District, British Columbia;* Wilson, 1913, *The Copper Smelting Industries of*

Canada, pp. 96, 109–110; and Fahey, 1994, *Shaping Spokane, Jay P. Graves and His Times*, p. 16.

[3] Minister of Mines, *Annual Report*, 1913.

[4] *The Mining Record*, June 1900, pp. 207–216.

[5] Wilson, 1913, *The Copper Smelting Industries of Canada*, pp. 109–111. There were many reports in *The Mining Record* including a lengthy description by C. A. Bramble, 1901, "The Boundary Creek District," pp. 207–221.

[6] It appears that the two railways cooperated in the use of this yard trackage and shared the use of a locomotive, probably each providing an engine for alternate periods, but this is not confirmed.

[7] Our description of the smelter is based largely on material in Wilson's 1913 study, *The Copper Smelting Industries of Canada*, pp. 120–132.

[8] Ibid., pp. 23–24.

[9] Ibid., especially pp. 120–132.

[10] Ibid., pp. 26–27.

[11] Lavallée, 1985, *Canadian Pacific Steam Locomotives*, pp. 79, 118, 279 and 334. See also the illustrated article by Al Welch, "Canadian Pacific Railway's Shay locomotive No. 111," including scale drawings, in *Railroad Model Craftsman*, April 2006, pp. 61–65. A letter from Thomas Shaughnessy to W. F. Tye, October 16, 1900, indicates the Shays were seen as a possible means of building cheaper, more direct but steeper branch lines in mining regions. Tye is instructed to survey a line with grades as steep as six or seven percent in the Lardeau region, north of Kootenay Lake, for possible use with Shay locomotives. CPR Shay No. 111 was featured in an article in *Scientific American*, "Freight Locomotive for Heavy Grades," Vol. 85, No. 5, August 24, 1901, p. 116. CPR Condition and Distribution of Locos reports for 1904 were provided by Omer Lavallée. The Lima builder's numbers and dates for the Shays were: 612, 8-9-00; 689, 4-16-02; and 824, 12-4-03, respectively.

[12] *The Phoenix Pioneer*, September 6, 1901, and *The Cascade Record*, September 22, 1901. Our account is based on these reports, but the damaged engine may have been left at Eholt where there were better repair facilities.

[13] *The Grand Forks News-Gazette*, August 27, 1904.

The Great Northern Expands and Builds to Phoenix, pages 112-127.

[1] J. J. Hill to Col. C. B. Lamborn, April 14, 1888, referring to a prospectus of the Crow's Nest Pass Coal Co. and to coal deposits near Great Falls; Hill to William Fertile, September, 1891; Hill to Van Horne, November 27, 1898.

[2] Hidy et al., 1988, *The Great Northern Railway*.

[3] Much of the route along the Kootenay was flooded by Lake Koocanusa in 1972 when the Libby Dam was built.

[4] See Eagle, 1989, *The Canadian Pacific and the Development of Western Canada*, pp. 115–122, for details of Hill's involvement with the Crow's Nest Pass Coal Company. See also *The Canadian Mining Journal*, October 1, 1907, p. 444.

[5] Hill to George A. Cox (Crow's Nest Pass Coal Co. president), January 14, 1907.

[6] *The Canadian Mining Journal*, September 1, 1907, p. 413.

[7] *The Morrissey Miner*, quoted in *The Grand Forks News-Gazette*, February 21, 1903.

[8] Hidy et al., 1988, *The Great Northern Railway*, pp. 133–135; *The Daily Colonist* (Victoria), April 14, 1907, p. 13.

[9] Canadian Pacific eventually purchased control of this line in 1916. Later the Union Pacific acquired a half interest in the railway and in 1958 took it over. It remains an important connection between Canadian Pacific and Union Pacific. See Turner, 1985, *West of the Great Divide*, pp. 149–150, and Fahey, 1965, *Inland Empire, D. C. Corbin and Spokane.*

[10] A set of the *Soo-Pacific's* passenger cars is being restored at the Canadian Museum of Rail Travel at Cranbrook.

[11] Quoted in *The Daily Colonist* (Victoria), July 20, 1907, from a *Wall Street Journal* article.

[12] *The Grand Forks News-Gazette*, June 29, 1904, and September 28, 1904.

[13] Ibid., June 4 and 11, 1904.

[14] Sanford, *McCulloch's Wonder*, p. 106.

[15] *Vancouver, Victoria & Eastern Railway and Navigation Co, Phoenix Line, summary account*, ca. 1905, GN President's subject file 4015, folder 16.

[16] Some records show the start of ore service being February 16, 1905.

[17] *The Grand Forks News-Gazette*, March 25, 1905.

[18] Ibid., February 28, March 11 and April 7, 1905.

[19] At this time Hill had control of the Northern Pacific as well as the Chicago, Burlington & Quincy, the latter usually called simply the "Burlington." Later Hill had to relinquish control of these railroads, but eventually, with the merger forming the Burlington Northern of March 2, 1970, his great vision of a larger system was realized, long after his death.

[20] *The Phoenix Pioneer*, September 30, 1905; *The Evening Sun* (Grand Forks), September 29, 1905; and *The Grand Forks Gazette*, August 5 and September 30, 1905. The list of guests is compiled from all three reports.

[21] Fahey, 1994, *Shaping Spokane, Jay P. Graves and His Times*, pp. 74–76.

[22] *The Grand Forks Gazette*, October 7, 1905.

[23] Fahey, 1994, *Shaping Spokane*, pp. 17 and 80–81; Eagle, 1989, *The Canadian Pacific Railway and the Development of Western Canada*; and Lamb, 1977, *History of the Canadian Pacific*, p. 236.Carloadings noted in this discussion and elsewhere are taken from notes compiled by Dave Wilkie from GN sources. *The Grand Forks Gazette*, May 6, 1905, noted that GN was carrying nearly twice as much ore as the CPR, but this may have been stockpiled ore awaiting the opening of the GN, and appears to be either exaggerated or of short-term duration. May's carloadings were 992 for the CPR and 743 for the GN. Daily trains, were bringing out 35 cars, each of 35 tons capacity, whereas the CPR trains were handling only 23 cars of 30 tons capacity.

[24] *The Grand Forks News-Gazette*, August 27, 1904.

[25] *The Grand Forks News-Gazette*, May 2, 1903, April 16, 1904, and April 30, 1904; *Anaconda News*, April 13, 1904.

[26] *The Grand Forks Gazette*, December 17, 1904.

[27] Ibid., January 14, 1905, and January 21, 1905.

[28] *The Phoenix Pioneer*, January 19, 1907.

[29] Ibid., February 9, 1907.

[30] *The Phoenix Pioneer*, January 19, June 8, and August 24, 1907, and September 18, 1909.

[31] *The Evening Sun* (Grand Forks), October 31, 1909. This wreck was also reported in *The Grand Forks Gazette*, November 4, 1909, *The Boundary Creek Times*, November 5, 1909, and *The Ledge* (Greenwood), November 4, 1909.

[32] This tragedy is recounted and discussed in considerable detail by George W. Pearce, 2003, "The Farron Hill Incident of 1909," *Branchline*, July–August, pp. 6–8.

Reciprocity, the Great War and the Boundary Railways, pages 128-133.

[1] Hill made many speeches supporting reciprocity. See *The Daily Colonist* (Victoria), August 25, 1908, p. 4, and November 28, 1908, p. 4.

[2] Martin, 1976, *James J. Hill and the Opening of the Northwest*, pp. 289–291 and 546–548.

[3] Eagle, 1989, *The Canadian Pacific Railway and the Development of Western Canada*, pp. 130–135.

[4] State of Washington, Public Service Commission of Washington, 1912, *Annual Report to the Governor*, and reports for 1913 and 1914.

[5] *Report of the Minister of Mines*, 1912, p. K174.

[6] *The Grand Forks Gazette*, November 4, 1911.

[7] Ibid., January 13, February 3 and 10, March 10, April 6 and May 25, 1912.

[8] LeRoy, 1912, *The Geology and Ore Deposits of Phoenix, Boundary District, British Columbia*, p. 22. See also Fahey, 1994, *Shaping Spokane, Jay P. Graves and His Times.*

[9] *The Grand Forks Gazette*, May 11, July 20 and October 5, 1912.

[10] Ibid., January 4 and 25, 1913.

[11] Ibid., August 3 and December 12, 1914, February 13, June 17 and October 23, 1915, and November 29, 1919; *The Phoenix Pioneer*, March 6 and May 8, 1915; *The Greenwood Ledge*, June 17, 1915.

Building Westward to the Similkameen, pages 134-153.

[1] See Sanford, 1977, *McCulloch's Wonder*, and 1990, *Steel Rails & Iron Men*; Smuin, 1997, *Canadian Pacific's Kettle Valley Railway* and 2003, *Kettle Valley Railway Mileboards, A Historical Field Guide to the KVR*; Turner's, 1995, *Steam on the Kettle Valley*; and Burrows, 1984, *Railway Mileposts British Columbia, Vol. II, The Southern Routes from the Crowsnest to the Coquihalla.*

[2] Cail, 1974, *Land, Man, and the Law,* pp. 165–168.

[3] Sometimes written Twentymile Creek.

[4] Camsell, 1910, *The Geology and Ore Deposits of Hedley Mining District, British Columbia*, p. 16.

[5] See Martin, 1976, *James J. Hill and the Opening of the Northwest,* pp. 348–349, and Malone, 1996, *James J. Hill, Empire Builder of the North West*, for background on Hill's relationship with Daly and the Montana copper industry.

[6] "Construction of the VV&E," 1905 clipping from *The Phoenix Pioneer and Boundary Mining Journal*, pp. 37–39.

[7] The line was described in detail in *The Boundary Creek Times*, August 24, 1906.

[8] *The Phoenix Pioneer*, February 25, 1906, and *The Boundary Creek Times*, March 9 and April 20, 1906.

[9] This track was removed in 1909.

[10] *The Grand Forks Gazette*, November 11, 1905, and a partial undated draft by Dave Wilkie; *The Boundary Creek Times*, October 27, 1905; and a 1905 clipping "Construction of the VV&E" and "Remarkable Railway Dispute," from *The Phoenix Pioneer and Boundary Mining Journal*, pp. 37–39. Estimates of the number of workers involved varied.

[11] *The Phoenix Pioneer*, July 27, 1906, and March 2, 1907. Also *The Boundary Creek Times*, September 7 and 14 and October 26, 1907.

[12] *The Phoenix Pioneer,* April 27, 1907, and *The Hedley Gazette*, April 25, 1907.

[13] *The Hedley Gazette*, April 4–18, 1907.

[14] Ibid., April 4–18, May 23 and July 11–August 1, 1907.

[15] Hidy et al., 1988, *The Great Northern Railway, A History,* pp. 134–135. The eight railway companies were the VV&E, Nelson & Fort Sheppard, Red Mountain, Crow's Nest Southern, Kaslo & Slocan, Kootenay Railway & Navigation, Kaslo & Lardo-Duncan and Bedlington & Nelson.

[16] *The Hedley Gazette*, July 18, 1907.

[17] *The Similkameen Star*, January 15, May 13 and August 12, 1908.

[18] See "Great Northern Covered Bridges," by Jack Porzig, *Reference Sheet No. 62*, Great Northern Railway Historical Society. This paper includes Great Northern drawings of 125-foot Howe truss bridges and many detailed photos of the bridges. The sheeting probably was applied between 1916 and 1920.

[19] Camsell, 1910, *The Geology and Ore Deposits of Hedley Mining District*, pp. 16–17.

[20] *The Similkameen Star,* July 14, October 13 and December 29, 1909, and January 5, 1910.

[21] Later Huntingdon–Sumas became an important gateway with rail services provided by Canadian Pacific, B.C. Electric (later B.C. Hydro and still later Southern Railway of British Columbia), as well as Northern Pacific (Burlington Northern) and the Milwaukee Road.

[22] A very helpful Australian website explaining gold mining terminology for this period is: www.ahc.gov.au/publications/generalpubs/mining/guideline6.html. Between 1988 and 1995 a large area of the mine tailings just east of Hedley were reworked in the Hedley Gold Tailings Project and some gold and silver was recovered. The tailings were agglomerated with cement and then leached with a cyanide solution to recover the metals and the solution was then recycled. Part of the project was to reclaim the worked-over tailings areas to reduce dust and protect water quality. The process used, in part, biosolids from the Greater Vancouver Regional District's wastewater treatment plant on Annacis Island and involved replanting with forage grasses. See Russ Horton and Roz Kempe, 2001, *Progressive Reclamation Utilizing Biosolids at the Hedley Gold Tailings Project, Hedley, British Columbia, Canada* at www.gvrd.bc.ca/nutrifor/pdfs/HedleyReclamation.pdf.

[23] Based on Camsell's 1910 study, *The Geology and Ore Deposits of Hedley Mining District*, pp. 180–184, and *B.C. Mining Review* and other mining journal reports, as well as Hedley Gold Mining Co. *Annual Reports* for 1912–1914, and the Hedley Mascot Gold Mines *Annual Report*, 1946, in the BC Archives. Also very helpful were Barnes' 1948 and 1950 papers on Hedley Camp and the Nickel Plate Mine, *British Columbia Historical Quarterly*. More recent is Cox, 1997, *Mines of the Eagle Country, Nickel Plate and Mascot,* which provides extensive background on these important mines.

[24] Hidy et al., 1988, *The Great Northern Railway*, pp. 320–324; Raymond Corley, 1967, ms. with later amendments, Turner collection.

[25] Statement of amounts expended to November 30, 1914, by the VV&E Ry & Navig. Co., GN President's subject file 4015.

[26] Martin, 1976, *James J. Hill and the Opening of the Pacific Northwest*, pp. 379–381.

[27] Turner, 1987, *West of the Great Divide.*

[28] *The Similkameen Star* carried frequent reports on the construction. See particularly May 25, 1910, and June 14, September 20 and November 15, 1911. The opening was reported in *CR&MW*, January 1913, p. 30.

[29] *CR&MW*, November 1912, p. 559. A formal agreement was completed on November 20, 1913, and expanded in a further agreement on July 10, 1914.

[30] Ibid.

[31] Sanford, 1990, *Steel Rails and Iron Men*, p. 54.

[32] Joe Smuin, personal communication, 2005, and James Mattson, 2001, *GN Ry Station Catalog*, Princeton to Hope, p. MA8-1.

[33] *Annual Report*, Minister of Mines, 1914, pp. K 364–K 368, K 475–K 482, and, for example, *Annual Report*, 1917, p. K491. It is clear that the GN did handle reasonable volumes of coal. As well as the report quoted on page 135, *The Penticton Herald* noted on November 8, 1922, that "the local coal company [at Princeton] had to suspend operations" due to a car shortage.

The GN's Third Main Line: Illusion and Reality, pages 154-161.

[1] *CR&MW,* May 1916, p. 182; *The Phoenix Pioneer*, March 11, 1911; Raymond Corley, 1967, ms. with later amendments, Turner collection.

[2] McCulloch, 1938, *The Railway Development in Southern British Columbia.*

[3] Warren to Budd, July 14, 1916. GN President's papers, 5072. Luthene C. Gilman retired from the GN as vice-president at Seattle in 1937.

[4] See Sanford, 1976, *McCulloch's Wonder*, chapters 10 and 11, and Eagle, 1989, *The Canadian Pacific and the Development of Western Canada*, pp. 136–137.

[5] Hidy et al., 1988, *The Great Northern Railway*, pp. 148–149.

[6] *The Penticton Herald*, June 29, 1916, and September 28, 1916, courtesy Joe Smuin. *The Greenwood Ledge*, September 1916.

[7] See, for example, Great Northern Railway timetable for November 1913. Also Ray Corley, 1967, ms. with later amendments, Turner collection.

[8] Malone, 1996, *James J. Hill, Empire Builder of the Northwest*, provides background on Hill's support for and interest in irrigation and dryland agriculture. See particularly pp. 251–252. See also Hidy et al., 1988, *The Great Northern Railway.*

[9] Hidy et al., 1988, *The Great Northern Railway*, pp. 150–152.

[10] Ibid.

[11] *CR&MW,* January 1914, p. 28, May 1916, p. 182 and February 1917 (article with artist's drawing); detailed original plans of the station courtesy Martin Evoy III; Grecula, 1993, "The Great Northern from Vancouver to the International Boundary," and "Terminal Developments at Vancouver, B.C.", *Railway Age Gazette,* November 23, 1917, Vol. 63, No. 21, pp. 945–949.

[12] Hidy et al., 1988, *The Great Northern Railway*, pp. 163–164; White, 1986, *The Great Yellow Fleet*, pp. 143–147.

[13] Hidy, ibid., pp. 325–327; Kenneth Askey, 2004, "The Railway to Port Guichon, 1903–1938," *The Sandhouse,* Autumn 2004, pp. 12–31; and Jones, "The Strategy of Railway Abandonment: The Great Northern in Washington and British Columbia, 1917–1935," *The Western History Quarterly,* April 1980.

[14] Statement of amounts expended to November 30, 1914, by the VV&E Railway & Navigation Co., GN President's subject file 4015. This accounting also includes in the total the cost of the Phoenix line and construction at Grand Forks and between Laurier and Danville.

Closure of the Phoenix and Motherlode Mines and the Boundary Smelters, pages 162-167.

[1] *The Grand Forks Gazette,* April 9, June 20 and 27, 1919, and *Granby News*, June 13, 1919.

[2] Jones, 1980, "The Strategy of Railway Abandonment: The Great Northern in Washington and British Columbia, 1917–1935." Budd's quotation is cited in this study and is from Budd to Kenney, April 24, 1919. GN President's Files, 8947.

[3] *The Greenwood Ledge*, May 15 and 22 and June 19, 1919; citations courtesy Joe Smuin. *Granby News*, 1919; Dave Wilkie notes; and CPR timetables.

[4] *The Grand Forks Gazette*, November 7 and 21, 1919.

[5] Ibid., April 9, 1919, and June 4 and August 6, 1920; tonnage reports from *Granby News*, December 1919.

[6] *The Greenwood Ledge*, July 15, 1926. Citation courtesy Joe Smuin.

[7] *The Grand Forks Gazette*, July 1 and 15, 1921.

[8] Ibid., July 15 and August 26, 1921, January 27, 1922, and August 17, 1923.

[9] Interstate Commerce Commission, *Valuation Docket No. 72, Spokane & British Columbia Railway Co., Submitted August 6, 1921. Decided March 12, 1926*, pp. 209–220.

[10] *The Grand Forks Gazette*, May 10 and 31, 1918, September 5 and 26, 1919, and July 8, 1923; Ray Corley, 1967, ms. with later amendments, Turner collection.

[11] Campbell, "The Passing of a Canadian City," *Canadian Mining Journal*, April 29, 1921, p. 334.

[12] Fahey, 1994, *Shaping Spokane, Jay P. Graves and His Times*, p. 82.

Great Northern Withdraws Along the Boundary, pages 168-179.

[1] E. T. Stansbury, undated, *Railroading* (memoirs in mimeographed publication *Railroads, The History of Great Northern Railways Line, The Spokane Falls & Northern*), Authors' collection, courtesy Mr. Stansbury.

[2] See the excellent discussion by Jones, "The Strategy of Railway Abandonment: The Great Northern in Washington and British Columbia, 1917–1935," *The Western Historical Quarterly*, April 1980, pp. 144–145.

[3] The statistics are from a Canadian Pacific diagram, 1927, *Comparative Profiles of K.V.R and C.P.R. Line with G.N.R. Line between Grand Forks and Princeton, B.C.*, CPR, Winnipeg. Courtesy of Barrie Sanford.

[4] These discussions are noted by Kennedy, 1986, *Canadian Pacific in Southern British Columbia, The Boundary Subdivision*, p. 5.

[5] Ibid., pp. 154–155.

[6] *The Penticton Herald*, March 7, 1935, and June 25 and July 2, 1936, courtesy Joe Smuin.

[7] See David C. Jones, 1981, "'Molson is Becoming Smaller': A Canadian-American Community After the First War," *BC Studies*, No. 49 (Spring), pp. 54–78. The photos of the route were made in 1934 as part of a survey of the line. The catalog numbers are: 85.52.1.2, and 1.8, 1.20, 2.14, 2.15, and 2.25.

[8] The GN acquired its first refrigerator cars in 1902, a venture that expanded into an extensive system, and in 1923 the operations were reorganized into the Western Fruit Express Co., a wholly owned GN subsidiary. This organization was integrated with the eastern-based Fruit Growers Express and soon after the Burlington's Refrigerator Express Co. The arrangement provided much greater efficiency and utilization of the refrigerator car fleet. See White, 1986, *The Great Yellow Fleet.* Barrie Sanford graciously shared his notes from *The Similkameen Star,* 1915–1918, and *The Princeton Star* from 1919–1926 on the ice harvest, see: November 11, 1905, March 1, 1918, January 17, 1919, February 25, 1921, February 2, 1923, January 18, 1924, and January 22, 1925. Joe Smuin's extensive GN notes from *The Penticton Herald* and other papers added further background.

[9] Joe Smuin, personal communication, 2005.

[10] *Memorandum of Agreements between Great Northern Railway, Vancouver Victoria & Eastern Railway & Navigation Company, Kettle Valley Railway Company, and Canadian Pacific Railway Company, in Connection with the Operation of Line between Hope and Brookmere.* Typescript of CPR file copy summary, December 30, 1948, Turner collection.

After the Boom Years, CPR Steam, pages 180-189.

[1] As part of the redress to Canadians of Japanese heritage, and as a memorial to this period, the Nikkei Internment Memorial Centre was established at New Denver on the site of one of the internment camps. It is a moving and beautifully developed heritage centre, with excellent exhibits and preserved buildings, and a Japanese garden designed by Mr. Roy Sumi when he was 87.

[2] Canadian Pacific Railway, 1952. *Application Study of Diesel-Electric Locomotives for the Kootenay and Kettle Valley Divisions*, CPR, Montreal, Quebec. Statistics used are based on information compiled in this report for train operations and summary figures provided in Appendix V, "Operating Data and Statistics." These difficult operating circumstances were confirmed by many conversations with operating crews and officials over the years.

[3] Ibid., "Assignment of Steam Locomotives," p. 72.

Diesel Years Along the Boundary, pages 190-207.

[1] See Keyes and Middleton, 1980, "The Great Northern Railway Company: All-Time Locomotive Roster, 1861–1970," *Railroad History*, No. 143 (reprinted 1989).

[2] Canadian Pacific Railway, 1952. *An Application Study of Diesel-Electric Locomotives for the Kootenay and Kettle Valley Divisions*, CPR, Montreal, Quebec. See also Dean and Hanna, 1981, *Canadian Pacific Diesel Locomotives*.

[3] Dynamic brakes make use of the locomotive's electric traction motors, which in braking mode are run as generators. The resistance of the motors acts as a brake and the heat generated from the system is dispelled by resister grids and cooling fans.

[4] Canadian Pacific Railway, 1952, *An Application Study of Diesel-Electric Locomotives for the Kootenay and Kettle Valley Divisions*.

[5] Kennedy, 1986, *Canadian Pacific in Southern British Columbia, The Boundary Subdivision*, p. 17.

[6] Turner, 1987, *West of the Great Divide*, pp. 267–272.

[7] See Canadian Transportation Agency, Decision 484-R-1990. Authorized for abandonment was trackage from Robson West (Mile 30.7) and Midway (Mile 16.6) as well as the Carmi Spur beyond Midway. See also the earlier initial ruling, Order No. 1988-R-547.

[8] See Canadian Transportation Agency, Decision 130-R-1993. The trackage conveyed was between miles 92.6 and 93.85 and included the Carson Spur in the city of Grand Forks.

[9] Canadian Transportation Agency, Order No. 1990-R-318.

[10] The Cascade & Columbia River RR operates over BNSF for a few miles into Wenatchee.

[11] By this time the BNSF was operating in British Columbia and Manitoba under its wholly owned subsidiary Burlington Northern (Manitoba) Ltd. In 1997, to reflect the merger with the Santa Fe, the subsidiary was renamed Burlington Northern Santa Fe Manitoba Inc. See Canadian Transportation Agency Decisions Nos. 403-R-1997 and 704-R-1997.

[12] See Canadian Transportation Agency, Decision No. 673-R-2004.

[13] OmniTRAX and Broe corporate websites, March 15, 2005. See US Surface Transportation Board Decision Docket AB9940X, July 6, 2006, for abandonment details, and Pope & Talbot, *Quarterly Report*, November 3, 2005.

Postscript: The Boundary Railways in Retrospect, pages 208-211.

[1] Jones, "The Strategy of Railway Abandonment: The Great Northern in Washington and British Columbia, 1917–1935," *The Western History Quarterly*, April 1980.

References

Many sources contributed to this book and specific references are indicated in the endnotes. More generally, the papers of the Great Northern Railway at the Minnesota Historical Society, the James J. Hill papers at the Hill Memorial Library in St. Paul, president's papers from Canadian Pacific's Archives, orders of the Board of Railway Commissioners, Department of Railways papers at the B.C. Archives, and the B.C. Department of Mines *Annual Reports* have been invaluable. Similarly helpful were Interstate Commerce Commission reports and dockets, and the *Annual Reports* of the Public Service Commission of Washington. We also consulted many local area newspapers (note that newspaper names often varied considerably over their years of publication) and they provided numerous details of the construction, operation and abandonment of the railways. Citations are included in the endnotes. *Canadian Railway and Marine World* (*CR&MW*) and its predecessor *The Railway and Shipping World* (*R&SW*) were excellent sources on the hectic construction period before the First World War as were mining industry journals such as *The British Columbia Mining Record* (*The Mining Record*).

We have also been fortunate in being able to interview a number of people who either worked for or travelled on the railways along the boundary, and their insights and comments provided a more personal view of the later years of operations. Similarly, early travel accounts have added a human side to the story that is important in understanding the roles played by the Great Northern and the Canadian Pacific in this region. Details of rights-of-way, mileages, facilities and similar information were drawn from engineering reports, railway site plans and maps as well as public and employee timetables. Several books expand on many of the themes and stories we have developed in this book. In particular: John Fahey's excellent biography of Daniel Corbin, *Inland Empire, D.C. Corbin and Spokane*; Albro Martin's fascinating benchmark biography of James J. Hill, *James J. Hill and the Opening of the Northwest;* John A. Eagle's thorough treatment of Canadian Pacific's role in western development, *The Canadian Pacific Railway and the Development of Western Canada,* which provides many insights into the corporate history of the CPR and the ongoing competition with the Great Northern; and Barrie Sanford's *McCulloch's Wonder, The Story of the Kettle Valley Railway* and its excellent companion volume, *Steel Rails and Iron Men, A Pictorial History of the Kettle Valley Railway,* give many insights into KVR construction and operation, competition between the railways and the accomplishments of Andrew McCulloch. An excellent guide with many maps and historical references is Roger Burrows' detailed *Railway Mileposts: British Columbia, The Southern Routes from the Crowsnest to the Coquihalla.* More recently Joe Smuin, with a long family background with the Kettle Valley, has produced an immensely detailed guide to the Kettle Valley Railway, *Kettle Valley Railway Mileboards.* Our earlier collaboration, *The Skyline Limited, The Kaslo and Slocan Railway,* presents the history of the Great Northern in the Kootenay Lake–Slocan mining district. The GN's operations were consolidated into the Kootenay Railway & Navigation Company that operated the rail and sternwheeler services from Bonners Ferry to Kootenay Lake, Kaslo and the Slocan. Two websites of particular interest are Donald M. Wilson's *The Virtual Crowsnest Highway* www.crowsnest-highway.ca, and the Canadian Museum of Rail Travel's *Crowsnest Pass Railway Route*, www.crowsnest.bc.ca, with text by Robert Turner.

Askey, Kenneth M. 2004. "The Railway to Port Guichon, 1903–1938," *The Sandhouse*, CRHA, Pacific Coast Division, Autumn 2004, pp. 12–31, and Winter 2004/05, "Riding the Guichon Line to Sumas and Back in 1924," pp. 8–27.

Barlee, N. L. (Bill). 1984. *Gold Creeks and Ghost Towns.* Hancock House, Surrey, BC.

Barnes, Harry D. 1948. "Early History of Hedley Camp," *British Columbia Historical Quarterly*, Vol. 12, No. 2, pp. 103–125.

———.1950. "The Nickel Plate Mine, 1898–1932," *British Columbia Historical Quarterly*, Vol. 14, No. 3, pp. 125–140.

Basque, Garnet. 1999. *Ghost Towns and Mining Camps of the Boundary Country*. Heritage House, Surrey, BC.

Bell, Archie. 1918. *Sunset Canada. British Columbia and Beyond.* The Page Company, Boston, MA.

Bramble, C. A. 1901. "The Boundary Creek District. The Town of Grand Forks and the Granby Smelter," *British Columbia Mining Record*, Vol. 7, No. 6, pp. 207–222.

Brown, Murray. 1988. *Gold at Nickel Plate*. Published by the Author, Richmond, BC.

Burrows, Roger. 1984. *Railway Mileposts: British Columbia, Vol. II, The Southern Routes from the Crowsnest to the Coquihalla.* Railway Milepost Books, North Vancouver, BC.

Cail, Robert E. 1974. *Land, Man and the Law, The Disposal of Crown Land in British Columbia, 1871–1913.* University of British Columbia Press, Vancouver, BC.

Campbell, C. M. 1902. "The Granby Mines at Phoenix," *The Mining Record*, Vol. 9, No. 11, pp. 367–372, and No. 12, pp. 401–405.

———. 1921. "The Passing of a Canadian City," *Canadian Mining Journal*, April 29, 1921, pp. 334–339.

Canada, Department of Mines. 1908. *Report on the Mining and Metallurgical Industries of Canada, 1907–08.* Ottawa, ON.

Canadian Pacific Railway. 1952. *Application Study of Diesel-Electric Locomotives for the Kootenay and Kettle Valley Divisions.* Research Department Report 30-52, Canadian Pacific Railway, Montreal.

Camsell, Charles. 1910. *The Geology and Ore Deposits of Hedley Mining District, British Columbia.* Memoir No. 2, Geological Survey, Canada Department of Mines, Ottawa, ON.

———. 1913. *Geology and Mineral Deposits of the Tulameen District, B.C.* Memoir No. 26, Geological Survey, Canada Department of Mines, Ottawa, ON.

Carter, L. E. 1979. *Granby, 75 Years of British Columbia Copper.* Graduating Essay, Simon Fraser University. Add Mss 1087, BC Archives, Victoria, BC.

Cheney, Eric S. and Michael G. Rasmussen. 1996. "Regional Geology of the Republic Area," *Washington Geology*, Vol. 24, No. 2, pp. 3–7.

Cousins, William James. 1981. *A History of the Crow's Nest Pass.* The Historic Trails Society of Alberta, Lethbridge, AB.

Cox, Doug. 1997. *Mines of the Eagle Country, Nickel Plate and Mascot.* Skookum Publications, Penticton, BC.

Cruise, David and Alison Griffiths. 1988. *Lords of the Line, The Men Who Built the CPR.* Viking, Markham, ON.

Dean, Murray W. and David B. Hanna. 1981. *Canadian Pacific Diesel Locomotives.* Railfare/Fitzhenry & Whiteside Ltd., Don Mills, ON.

Dempsey, Hugh (editor). 1984. *The CPR West, The Iron Road and the Making of a Nation.* Douglas & McIntyre, Vancouver, BC.

Doeksen, Gerry. 1985. *Railways of the Boundary.* Railways of Western Canada, Vol. 4. Published by the Author, Montrose, BC.

Doeksen, Gerry and Corwin. 1988. *Railways of the West Kootenay, Part 3.* Railways of Western Canada, Vol. 5. Published by the Authors, Montrose, BC.

Dorman, Robert. 1938. *A Statutory History of Steam and Electric Railways of Canada, 1836–1937.* Canada Department of Transport, Ottawa, ON.

Eagle, John A. 1989. *The Canadian Pacific Railway and the Development of Western Canada.* McGill-Queens University Press, Kingston, ON.

Ewert, Henry. 1986. *The Story of the B.C. Electric Railway Company.* Whitecap Books, North Vancouver, BC.

———. 2005. "Great Northern's Third Route to the Pacific,

Spokane to Vancouver, B.C." *Reference Sheet No. 325,* Great Northern Railway Historical Society, St. Paul, MN.

Fahey, John. 1965. *Inland Empire, D.C. Corbin and Spokane.* University of Washington Press, Seattle, WA.

———. 1986. *The Inland Empire, Unfolding Years, 1879–1929.* University of Washington Press, Seattle, WA.

———. 1994. *Shaping Spokane, Jay P. Graves and His Times.* University of Washington Press, Seattle, WA.

Findlay, John M. and Ken S. Coates (editors.) 2002. *Parallel Destinies, Canadian-American Relations West of the Rockies.* McGill-Queens University Press, Montreal, PQ, and Kingston, ON.

Fleming, R. B. 1991. *The Railway King of Canada, Sir William Mackenzie 1849–1923.* University of British Columbia Press, Vancouver, BC.

Fowler, S. S. 1939. "Early Smelters in British Columbia," *British Columbia Historical Quarterly,* Vol. 3, No. 3, pp. 183–201.

Garden, John F. 2004. *The Crow and the Kettle.* Footprint Publishing, Cowley, AB.

Glanville, Jim and Alice. 1997. *The Life and Times of Grand Forks, Where the Kettle River Flows.* Blue Moose Publications, Kelowna, BC.

Grecula, Walter A. 1993. "The Great Northern from Vancouver to the International Boundary."*Reference Sheet No. 200,* Great Northern Railway Historical Society, St. Paul, MN.

Green, Mervyn. 1987. *British Columbia Industrial Locomotives, An All-Time Listing, Including Short Lines.* Published by the Author, Richmond, BC.

Greever, William S. 1963. *Bonanza West, The Story of the Western Mining Rushes, 1848–1900.* University of Idaho Press, Moscow, ID.

Hall, Harold R. 1998. *Stations and Depots of the Great Northern Railway.* Manuscript courtesy of the author.

Harris, Cole. 1983. "Moving Amid the Mountains," *BC Studies,* No. 58, pp. 3–39.

———. 1997. *The Resettlement of British Columbia.* University of British Columbia Press, Vancouver, BC.

Hearn, George and David Wilkie. 1966. *The Cordwood Limited, A History of the Victoria and Sidney Railway.* British Columbia Railway Historical Association, Victoria, BC.

Hedges, James B. 1939 (1971 edition). *Building the Canadian West, The Land and Colonization Policies of the Canadian Pacific Railway.* Russell & Russell, New York, NY.

Hidy, Ralph W., Muriel E. Hidy and Roy V. Scott with Don L. Hofsommer. 1988. *The Great Northern Railway, A History.* Harvard Business School Press, Boston, MA.

Hill, Beth. 1989. *Exploring the Kettle Valley.* Polestar Press, Winlaw, BC.

Hodges, L. K. 1897. *Mining in the Pacific Northwest.* The Post-Intelligencer, Seattle, WA. (Reprinted by Shorey Book Store, Seattle, WA, 1970.)

Jones, David C. 1980. "The Strategy of Railway Abandonment: The Great Northern in Washington and British Columbia, 1917–1935," *The Western Historical Quarterly,* Vol. 11 (April), pp. 141–158.

———. 1981. "'Molson Is Becoming Smaller': A Canadian-American Community After the First War," *BC Studies,* No. 49 (Spring), pp. 54–78.

Kennedy, W. Gibson. 1983. *Canadian Pacific's Rossland Subdivision.* British Railway Modellers of North America, Calgary, AB.

———. 1986. *Canadian Pacific in Southern British Columbia, The Boundary Subdivision.* British Railway Modellers of North America, Calgary, AB.

Keyes, Norman C., Jr., and Kenneth R. Middleton. 1980. "The Great Northern Railway Company: All-Time Roster, 1861–1970," *Railroad History,* No. 143 (Autumn), pp. 20–162. (Reprinted by the Great Northern Railway Historical Society, 1989.)

Kirk, Ruth and Carmela Alexander. 1990. *Exploring Washington's Past, A Road Guide to History.* University of Washington Press, Seattle, WA.

Klein, Murray. 1990. *The Union Pacific, The Rebirth, 1894–1969.* Doubleday, New York, NY.

———. 2000. *The Life and Legend of E. H. Harriman.* University of North Carolina Press, Chapel Hill, NC.

Knowles, Valerie. 2004. *From Telegrapher to Titan, The Life of William C. Van Horne.* The Dundurn Group, Toronto, ON.

Lamb, W. Kaye. 1977. *History of the Canadian Pacific Railway.* Macmillan, New York, NY.

Langford, Dan and Sandra. 1997. *Cycling the Kettle Valley Railway* (2nd edition). Rocky Mountain Books, Calgary, AB.

LaPierre, Laurier L. 1996. *Sir Wilfrid Laurier and the Romance of Canada.* Stoddart Publishing, Toronto, ON.

Larson, Andrew G. and Clarence S. Verrill. 1915. *The Mineral and Other Resources of the North Fork of the Kettle River in the Grand Forks Mining District.* Bulletin No. 3, 1914, British Columbia Bureau of Mines, Victoria, BC.

Lasmanis, Raymond. 1996. "A Historical Perspective on Ore Formation Concepts, Republic Mining District, Ferry County, Washington," *Washington Geology,* Vol. 24, No. 2, pp. 8–14.

Lavallée, Omer S. A. 1974. *Van Horne's Road.* Railfare, Toronto, ON.

———. 1985. *Canadian Pacific Steam Locomotives.* Railfare, Toronto, ON.

Leonard, Frank. 1996. *A Thousand Blunders, The Grand Trunk Pacific Railway and Northern British Columbia.* University of British Columbia Press, Vancouver, BC.

LeRoy, O. E. 1912. *The Geology and Ore Deposits of Phoenix, Boundary District, British Columbia.* Memoir No. 21, Geological Survey, Canada Department of Mines, Ottawa, ON.

———. 1913. *Mother Lode and Sunset Mines, Boundary District, B.C.* Memoir 19, Geological Survey, Canada Department of Mines, Ottawa, ON.

Lindgren, Waldemar and Howland Bancroft. 1914. *The Ore Deposits of Northeastern Washington, Including a Section on The Republic Mining District.* United States Geological Survey, Bulletin 550, Washington, DC.

McCulloch, Andrew. 1938. *The Railway Development in Southern British Columbia from 1890 on, and Some Reasons for Building the Kettle Valley Railway, and the Line of the Vancouver, Victoria & Eastern Railway & Navigation Company in British Columbia.* Typescript, November 1938. Turner collection.

McNaughton, D. A. 1945. *Greenwood–Phoenix Area British Columbia.* Paper 45-20, Canada Department of Mines and Technical Surveys, Ottawa, ON.

McNelis, Sarah. 1968. *Copper King at War, The Biography of F. Augustus Heinze.* University of Montana Press, Missoula, MT.

Malone, Michael P. 1996. *James J. Hill, Empire Builder of the Northwest.* University of Oklahoma Press, Norman, OK.

Marcosson, Isaac F. 1957. *Anaconda.* Dodd, Mead & Co., New York, NY.

Martin, Albro. 1976. *James J. Hill and the Opening of the Northwest.* Oxford University Press, New York, NY.

———. 1991. "Hill or Harriman—What Difference Did It Make to Spokane?" in Stratton, David H. (editor), *Spokane and the Inland Empire, An Interior Pacific Northwest Anthology,* Washington State University Press, Pullman, WA, pp. 109–122.

Mattson, James C. 2004. *Great Northern Railway Mileposts, Station Catalog.* Manuscript produced by the author, Seattle, WA.

Meinig, D. W. 1998. *The Shaping of America, A Geographical Perspective on 500 Years of History, Vol. 3, Transcontinental America, 1850–1915.* Yale University Press, New Haven, CT.

Mouat, Jeremy. 1995. *Roaring Days, Rossland's Mines and the History of British Columbia.* University of British Columbia Press, Vancouver, BC.

———. 2002. "Nationalist Narratives and Regional Realities: The Political Economy of Railway Development in Southeastern British Columbia, 1895–1905," in Findlay, John M. and Ken S.

Coates (editors), *Parallel Destinies, Canadian-American Relations West of the Rockies,* McGill-Queens University Press, Montreal, QC and Kingston, ON, pp. 123–151.

Muralt, Darryl E. 1985. *Steel Rails and Silver Dreams, A History of the Dolly Varden Mines and the Narrow Gauge Dolly Varden Mines Railway.* Benchmark Publishing, Los Altos, CA.

———. 1992. *V.& S., The Victoria and Sidney Railway.* British Columbia Railway Historical Association, Victoria, BC.

Mussio Ventures and Trails BC. 2001. *Trans Canada Trail, the British Columbia Route.* Mussio Ventures, New Westminster, BC.

Peters, Edward Dyer, Jr. 1895. *Modern Copper Smelting* (7th edition). The Scientific Publishing Company, New York, NY, and London, England.

Porzig, Jack. 1988. "Great Northern Covered Bridges." *Reference Sheet No. 62,* Great Northern Railway Historical Society, St. Paul, MN.

Regehr, T. D. 1976. *The Canadian Northern Railway.* Macmillan of Canada, Toronto, ON.

Riegger, Hal. 1981. *The Kettle Valley and Its Railways.* PFM Publications, Edmonds, WA.

Roy, Patricia E. 1980. "Progress, Prosperity and Politics: The Railway Politics of Richard McBride,"*BC Studies,* No. 47, pp. 3–28.

Sanford, Barrie. 1977. *McCulloch's Wonder, The Story of the Kettle Valley Railway.* Whitecap Books, West Vancouver, BC.

———. 1990. *Steel Rails and Iron Men, A Pictorial History of the Kettle Valley Railway.* Whitecap Books, Vancouver, BC.

———. 2004. *Royal Metal, The People, Times and Trains of New Westminster Bridge.* National Railway Historical Society, British Columbia Chapter, Vancouver, BC.

Schnabel, Carl. 1898. *Handbook of Metallurgy, Vol. 1, Copper-Lead-Silver-Gold.* MacMillan & Co., New York, NY.

Schwantes, Carlos A. 1989. *The Pacific Northwest, An Interpretive History.* University of Nebraska Press, Lincoln, NB.

———. 1993. *Railroad Signatures Across the Pacific Northwest.* University of Washington Press, Seattle, WA.

———. 1999. *A Long Day's Journey, The Steamboat and Stage Coach Era in the Northern West.* University of Washington Press, Seattle, WA.

Smuin, Joe. 1997. *Canadian Pacific's Kettle Valley Railway.* British Railway Modellers of North America, Calgary, AB.

———. 2003. *Kettle Valley Railway Mileboards, A Historical Field Guide to the KVR.* North Kildonan Publications, Winnipeg, MB.

Stratton, David H. (editor). 1991. *Spokane and the Inland Empire, An Interior Pacific Northwest Anthology.* Washington State University Press, Pullman, WA.

Stuckey, Lawrence A. and Donald M. Bain. 1996. *The Great Northern and Northern Pacific Railways in Canada, Vol. 1.* British Railway Modellers of North America, Calgary, AB.

Taylor, Geoffrey W. 1988. *The Railway Contractors.* Morriss Publishing, Victoria, BC.

Trimble, William Joseph. 1914. *The Mining Advance into the Inland Empire.* Bulletin of the University of Wisconsin, No. 638, Madison, WI. (Reprinted 1986 by Ye Galleon Press, Fairfield, WA.)

Turner, Robert D. 1984 and 1998. *Sternwheelers and Steam Tugs, An Illustrated History of the Canadian Pacific Railway's British Columbia Lake and River Service.* Sono Nis Press, Victoria, BC.

———. 1987. *West of the Great Divide, An Illustrated History of the Canadian Pacific Railway in British Columbia, 1880–1986.* Sono Nis Press, Victoria, BC.

———. 1991. *The SS Moyie, Memories of the Oldest Sternwheeler.* Sono Nis Press, Victoria, BC.

———.1995. *Steam on the Kettle Valley, A Railway Heritage Remembered.* Sono Nis Press, Victoria, BC.

Turner, Robert D. and David S. Wilkie. 1994. *The Skyline Limited, The Kaslo and Slocan Railway.* Sono Nis Press, Victoria, BC.

Umpleby, Joseph B. 1910. *Geology and Ore Deposits of Republic Mining District.* Washington Geological Survey, Bulletin No. 1, Olympia, WA.

Veazey, Phyllis. 1983. "John Hendry and the Vancouver, Westminster and Yukon Railway: 'It Would Put Us on Easy Street,'" *BC Studies,* No. 59, pp. 44–63.

Walter, Edward M. and Susan A. Fleury. 1985. *Eureka Gulch; the Rush for Gold, A History of Republic Mining Camp 1896–1908.* The Authors, Colville, WA.

White, John H. 1986. *The Great Yellow Fleet, A History of American Railroad Refrigerator Cars.* Golden West Books, San Marino, CA.

Wilson, Alfred W. G. 1913. *The Copper Smelting Industries of Canada.* Mines Branch (No. 209), Canada Department of Mines, Government Printing Bureau, Ottawa, ON.

Wood, Charles and Dorothy. 1979. *The Great Northern Railway, A Pictorial Study.* Pacific Fast Mail, Edmonds, WA.

Woodcock, George and Ivan Avakumovic. 1968. *The Doukhobors.* Oxford University Press, Toronto, ON.

Zuters, Gary. 1994. *CP Rail 1993 Review.* Hyrail Productions, Ferndale, WA.

The view from the train high above Lower Arrow Lake was dramatic, even more so on a snowy winter day in the late 1930s. The lower photo, also from the late 1930s, presents a fireman's view along the boiler of a 3600 series 2-8-0 at Castlegar.
–L. S. Morrison, both photos.

Index

References in italics indicate illustrations. Additional illustrations may be within the page ranges. Mines, claims and geographic features are in British Columbia unless noted. See also the listing of railway names and abbreviations on pages 21-23.

Gib Kennedy captured the rugged beauty and excitement of travel through the Boundary District in this portrait of Canadian Pacific's Train No. 12, *The Kootenay Express*, near Fife, B.C., on its westward way through the mountains to Grand Forks, Penticton and Vancouver, in October 1944.